BIOCHEMISTRY RESEARCH TRENDS

CHEMICAL AND BIOCHEMICAL KINETICS

NEW PERSPECTIVES

BIOCHEMISTRY RESEARCH TRENDS

Additional books in this series can be found on Nova's website under the Series tab.

Additional E-books in this series can be found on Nova's website under the E-books tab.

BIOCHEMISTRY RESEARCH TRENDS

CHEMICAL AND BIOCHEMICAL KINETICS

NEW PERSPECTIVES

GENNADY ZAIKOV
EDITOR

Nova Science Publishers, Inc.
New York

Library of Congress Cataloging-in-Publication Data

Chemical and biochemical kinetics : new perspectives / editors, Gennady Zaikov.
p. cm.
Includes index.
ISBN 978-1-61209-224-9 (softcover)
1. Chemical kinetics. I. Zaikov, G. E. (Gennadii Efremovich), 1935-
QD502.C4654 2011
541'.394--dc22

2010054274

Published by Nova Science Publishers, Inc. † New York

This volume is dedicated to the memory of Frank Columbus

On December 1st 2010, Frank H. Columbus Jr. (President and Editor-in-Chief of Nova Science Publishers, New York) passed away suddenly at his home in New York.

We lost our colleague, our good friend, a nearly perfect person who helped scientists from all over the world. Particularly Frank did much for the popularization of Russian and Georgian scientific research, publishing a few thousand books based on the research of Soviet (Russian, Georgian, Ukranian etc.) scientists.

Frank was born on February 26th 1941 in Pennsylvania. He joined the army upon graduation of high school and went on to complete his education at the University of Maryland and at George Washington University. In 1969, he became the Vice-President of Cambridge Scientific. In 1975, he was invited to work for Plenum Publishing where he was the Vice-President until 1985, when he founded Nova Science Publishers, Inc.

Frank Columbus did a lot for the prosperity of many Soviet (Russian, Georgian, Ukranian, Armenian, Kazakh, Kyrgiz, etc.) scientists publishing books with achievements of their research. He did the same for scientists from East Europe – Poland, Hungary, Czeckoslovakia (today it is Czeck republic and Slovakia), Romania and Bulgaria.

He was a unique person who enjoyed studying throughout the course of his life, who felt at home in his country which he loved and was proud of, as well as in Russia and Georgia.

There is a famous Russian proverb: "The man is alive if people remember him." In this case, Frank is alive and will always be in our memories while we are living. He will be remembered for his talent, professionalism, brilliant ideas and above all – for his heart.

Professor Gennady Efremovich Zaikov
Honoured Member of Russian Science
Head of Polymer Division, IBCP

CONTENTS

PREFACE

"The aim of Science is to lead our mind
following the Creator's reason"
Isaac Newton

Some chapters with information about new steps of success in the field of chemical and biochemical kinetics have been accumulated in this volume.

Birth of kinetics as a science is 1881 when two Dutch scientists (Guldberg and Waage) published the paper about the Law of Mass Actions.

If we have chemical reactions between two molecules (A and B) A+B---$^k{}_2$---C the rate of reaction is equal $W_2 = k_2$ [A][B] where W_2 is the rate of reaction, k_2 – constant of rate of reaction, [A] and [B] –the concentration of A and B.

The second step in this part of science was developed by Swedish scientist Swante Arrenius (1889). He explained for scientists the meaning of Energy of activation of reaction E

$$K_2 = A_2 \cdot \exp(-E/RT) \text{ l/mol}\cdot\text{sec}$$

where A_2 is the predexponentional factor

In case of monomolecular reaction:

A-B----$^k{}_1$-----C+D $W_1 = k_1[AB]$ and $k_1 = 10^{13}\exp(-E/RT)\sec^{-1}$ where 10^{13} is amount of bond's vibration in molecule A-B during one second.

The readers can find the following information in this volume: activation energy of molecule and atom diffusion in bio-systems, mathematical modeling dependence of thermodynamic characteristics upon spatial-energy parameter of free atoms, dependence of nanostructure formation degree upon spatial-energy characteristics of free atoms, sevilene-based hybrid composites, organosilicon polymers with photo switchable fragments in the side chain, ferrocene-containing polyphenylenes as precursors for magnetic nanomaterials, moisture permeability and swelling of modified cellulose diacetate films, relation between the cell-free DNA content and the lipid peroxidation in the blood plasma of mice under damaging action, quantum-chemical calculations of electron structure of heme and ion of iron and their reactivity in reactions with hydrogen peroxide, effect of chitosan molecular weight on rheological behavior of chitosan modified nanoclay in highly hydrated states, the structure of polybutadienes and butadiene-acrylonitrile copolymers, inhibition mechanism of natural and synthetic antioxidants in the presence of phospholipids and hydroperoxides, physicochemical and antimicrobial properties of nanocomposite films based on hydroxyethylcellulose, electric conductivity of polymer composites at mechanical relaxation, quantum-chemical research of the mechanism of synthesis of 5-acetyloxymethyl-2-chloro-5-ethyl-1,2,3-dioxaphosphorynane, estimation of acid force of components of synthesis of 5-acetyloxymethyl-2-chloro-5-ethyl-1,2,3-dioxaphosphorynane, structure and properties of polymer adsorption layers on inorganic pigments surface in aqueous dispersion formed under ultrasonic treatment, pigment surface modification in pigment dispersions with thermo-responsive poly(methylvinylether) copolymers in combination with ultrasonic treatment, thermophysical properties of organoplastics based on PA-6, IR-spectra and a crystalline structure of penton and the carboplastic based on it, mechanism of stable radical generation in lignin under the action of nitrogen dioxide, properties of carbon-black composite vapor detectors based on multifunctional polymers, template synthesis of bioinorganic polymers on ions of transitive metals.

The editor is waiting for comments of readers of this book. Contributors and editor will be happy to take in account the comments in our research in future.

In: Chemical and Biochemical Kinetics: New Perspectives
Editor: Gennady Zaikov
ISBN 978-1-61209-224-9

Chapter 1

Activation Energy of Molecule and Atom Diffusion in Bio-systems

G. A. Korablev,[1,*] and G. E. Zaikov[2,#]

[1]Basic Research-educational Center of Chemical Physics and Mesoscopy, UdRC UrD, Russian Academy of Sciences, Izhevsk, Russia

[2]N. M. Emanuel Institute of Biochemical Physics, Russian Academy of Sciences, Moscow, Russia

Abstract

A methodology for estimating activation energy of diffusion processes in bio-membranes was developed based on spatial-energy notions (P-parameter). It is shown that transport relative penetrability in bio-membranes numerically equals to the degree of conformation interactions calculated via the relative difference of corresponding P-parameters. Calculation criteria are in accordance with reference and experimental data.

Keywords: spatial-energy parameter, diffusion, activation energy, bio-structures, relative penetrability

Aims and Background

Transport (transition) of molecules and atoms via bio-membranes is usually classified by energy-dependence [1]. Active transport occurs at the consumption of chemical energy due to hydrolysis of ATP and, strictly speaking, is not a diffusion process. Passive transport, i. e.the transition of a substance by the gradient of electrochemical potential, is not connected with the consumption of chemical energy by the cell: this is a diffusion process to the direction of the least electrochemical potential. The availability of trans-membrane gradient of molecule and atom concentration is its electromotive force.

* For correspondence: E-mail: korablev@udm.net.
# E-mail:chembio@chph.ras.ru.

The following types of passive transport are distinguished:

(i) simple diffusion of molecules conforming to the well-known Fick's law;
(ii) diffusion through the pores (channels), selective penetrability of which can be limited by the solubility of substances in lipids;
(iii) Transport with the help of molecules-carriers that can be executed in two variants:

(a) mobile carrier – diffusion of molecules takes place alongside with a substance in the membrane;
(b) forwarding of a substance from one molecule to another.

Besides, the diffusion can proceed not only through the membrane, but also through movable water layers adjacent to the membrane from both sides, for example, oxygen diffusion into the cell. In these cases, the total system penetrabilityis usually estimated based on the principle of adding inverse values of penetrability in each subsystem:

$$\frac{1}{\gamma}=\frac{1}{\gamma_M}+\frac{1}{\gamma'_W}+\frac{1}{\gamma''_W} \quad (1)$$

The effective penetrability (γ) is proportional [2] to the diffusion coefficient (D), membrane square (S), channel radius (r):

$$\gamma=\pi r^2 S\frac{D}{l} \quad (1a)$$

where l is thickness of the membrane.

In a more modern model of limited diffusion [3, 4] the protein fragment is considered as the Brownian oscillator with great attenuation (K. V. Shaytan, A. B. Rubin). Its motion is described by two parameters: conformation potential and diffusion coefficient depending upon the coordinate x, which results in the general equation of the fragment movement in dense medium (the Langevene equation).

In electro-diffusion processes the energy of ion penetration into the non-polar phase (W) is defined by the Born formula [4]:

$$W=\left[\frac{Z^2e^2}{2r}\right]\left(\frac{1}{\varepsilon_W}-\frac{1}{\varepsilon_M}\right) \quad (2)$$

where Z isthe ion valence, e – elementary charge, r – ion radius, ε_W and ε_M –dielectric penetrability of water and membrane.

It is interesting to find the direct dependence of main parameters of membrane diffusion with initial spatial-energy characteristics of free atoms. We showed such a dependence before on the examples of diffusion and self-diffusion of atoms in many simple and complex structures [5, 6].

P-parameter

The analysis of kinetics of different physical and chemical processes shows that in many cases inverse values of velocities, kinetic or energy characteristics of corresponding interactions are added.

Some examples: ambipolar diffusion, total rate of topochemical reaction, change in the light velocity during the transition from vacuum into the given medium.

In particular, such a supposition is proved by the aforesaid equations (1) and (2), and also by the formula of probability of electron transport process (W_∞) due to the overlapping of functions 1 and 2 (in stationary state) during electron-conformation interactions:

$$W_\infty = \frac{1}{2}\frac{W_1 W_2}{(W_1 + W_2)} \text{ or } \frac{1}{2W_\infty} = \frac{1}{W_1} + \frac{1}{W_2} \qquad (3)$$

Equation (3) is used [3] when estimating the characteristics of diffusion processes followed by radiationless transitions of electrons in proteins.

The Lagrange equation can serve as an example. For relative motion of isolated system of two interacting material points with masses m_1 and $m_2$in coordinate x it looks as follows:

$$m_{eq} x'' = -\frac{\partial U}{\partial x}$$

where $\frac{1}{m_{eq}} = \frac{1}{m_1} + \frac{1}{m_2}$

and U – mutual potential energy of points; m_{eq} – equivalent mass. At the same time $x''=a$ (characteristic of system acceleration). For elementary sections of interactions Δx can be accepted as:

$$\frac{\partial U}{\partial x} \approx \frac{\Delta U}{\Delta x} \text{ i.e.: } m_{np} a \Delta x = -\Delta U .$$

Then:

$$\frac{1}{1/(a\Delta x)} \frac{1}{(1/m_1 + 1/m_2)} \approx -\Delta U ;$$

$$\frac{1}{1/(m_1 a\Delta x) + 1/(m_2 a\Delta x)} \approx -\Delta U$$

or: $$\frac{1}{\Delta U} \approx \frac{1}{\Delta U_1} + \frac{1}{\Delta U_2} \qquad (4)$$

where ΔU_1 and ΔU_2 are potential energies of material points on elementary section of interactions, ΔU – resultant (mutual) potential energy of these interactions.

The atom system is formed from differently charged masses of nucleus and electrons. In this system energy characteristics of subsystems are as follows: orbital energy of electrons and effective energy of nucleus taking into account screening effects (by Clement).

Therefore, assuming that the resultant interaction energy of the system orbital-nucleus (responsible for interatomic interactions) can be calculated by the principle of adding inverse values of some initial energy components, we substantiate the introduction of P-parameter[6] as the averaged characteristic of valence orbitals is based on the following equations:

$$\frac{1}{q^2/r_i} + \frac{1}{W_i n_i} = \frac{1}{P_E} \qquad (5)$$

$$P_E = \frac{P_0}{r_i} \qquad (6)$$

$$\frac{1}{P_0} = \frac{1}{q^2} + \frac{1}{(Wrn)_i} \qquad (7)$$

$$q = \frac{Z^*}{n^*} \qquad (8)$$

where W_i is the bond energy of electrons [7] orionisation energy of atom (E_i) (Ref. 8); r_i– orbital radius$i$–orbital [9]; n_i – number of electrons of a given orbital, Z^* and n^* – effective charge of nucleus and effective main quantum number [10].

P_0 will be called spatial-energy parameter, and P_E – effective P-parameter. Effective P_E-parameter has a physical sense similar to some averaged energy of valence electrons inside the atom and is measured in energy units, for example, in electron-volts (eV).

Now we will briefly discuss the reliability of such an approach. According to the calculations[6] the values of P_E-parameters are numerically equal (in the range of 2%) to the total energy of valence electrons (U) by statistic model of the atom. Using a well-known relation between electron density (β) and interatomic potential, we can obtain the direct dependence of P_E-parameter upon the electron density at the distance $r_i$from the nucleus by statistic model of the atom:

$$\beta_i^{2/3} = \frac{AP_0}{r_i} = AP_E$$

whereAisaconstant.

The rationality of such an equation was proved by the calculation of electron density using the Clementi wave functions [10] and its comparison with the value of electron density calculated through the value of P_E-parameter.

The correlations of modules of maximum values of radial part of Ψ-function with the values of P_0-parameter were carried out and linear dependence between these values was found. Using some properties of wave function as applied to P-parameter, we obtain the wave equation of P-parameter formally analogous to the equation of Ψ-function.

Based on the calculations and comparisons, two principles of adding spatial-energy criteria depending upon wave properties of P-parameter and systemic character of interactions and charges of particles are substantiated:

1. Interaction of differently charged (heterogeneous) systems consisting of I, II, III, ... grades of atoms is satisfactorily described by the principle of adding the corresponding inverse energy values based onequations (5) – (8), thus corresponding to the minimum attenuation of oscillations taking place in anti-phase,
2. During the interaction of similarly charged (homogeneous) subsystems the principle of algebraic addition of their P-parameters by the following equations is fulfilled:

$$\sum_{i=1}^{m} P_0 = P_0^{'} + P_0^{''} + \ldots + P_0^{m} \qquad (9)$$

$$\sum P_Э = \frac{\sum P_0}{R} \qquad (9a)$$

where R is a dimensional characteristic of atom (or chemical bond) – this corresponds to the maximum intensification of oscillations taking place in the phase.

Applying the methodology of P-parameter based on the experimental data of about a thousand of different systems, spatial-energy principles of isomorphic replacement were found:

1. Complete (100%) isomorphic replacement and structural interaction occur at approximate equality of P-parameters of valence orbitals of interacting atoms: $P_E^{'} \approx P_E^{''}$
2. P-parameters of valence orbital of the atom of the least value define the orbital that is mainly responsible for the isomorphism and structural interactions.

Modifying the rules of adding inverse values of energy magnitudes of subsystems as applied to complex structures, we can propose the equation for calculating the P_C-parameter of a complex structure [6]:

$$\frac{1}{P_S} = \left(\frac{1}{NP_E}\right)_1 + \left(\frac{1}{NP_E}\right)_2 + \ldots \quad (10)$$

where N_1 and N_2 arenumbers of homogeneous atoms in subsystems.

During the formation of solution and other structural interactions in the areas of the contact of atoms-components, the same electron density has to be established. This process is followed by the redistribution of electron density between valence areas of both particles and transition of the part of electrons from external spheres into the neighbouring ones.

Apparently, with the nearness of electron densities in free atoms-components, the transition processes between boundary atoms of particles will be minimal, thus favouring the formation of a new structure. Thus, the task for estimating the degree of structural interactions, in many cases, comes to the comparative evaluation of electron density of valence electrons in free atoms (on the averaged orbitals) participating in the process based on the following equation:

$$\alpha = \frac{P'_E - P''_E}{P'_E + P''_E} 200\% \quad (11)$$

Following this methodology the degree of structural interactions (ρ) was evaluated in many (over a thousand) of simple and complex systems [6]. The nomogram of the dependence of ρ on the coefficient of structural interaction (α) (for P-parameters calculated via the ionisation energy) was obtained (Figure 1).

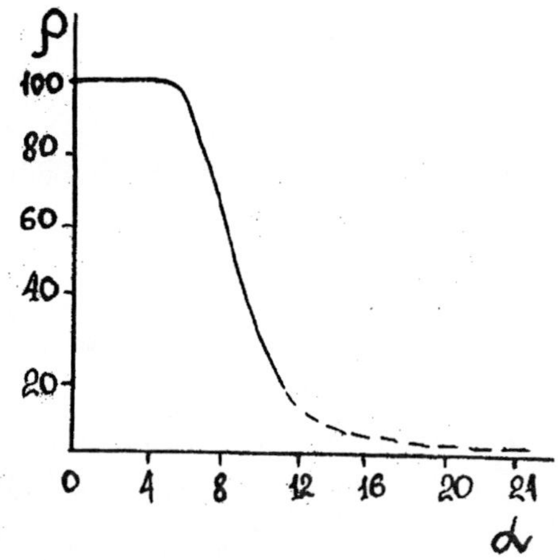

Figure 1. Nomogram of the dependence of the degree of structural interaction (ρ) on the coefficient α – for P-parameters calculated via the ionisation energy (E_i).

Isomorphism as a phenomenon usually is applied to crystalline structures. But obviously, the analogous processes can also take place between molecular compounds where their role and importance are not worse than of purely coulomb interactions.Diffusion processes, in particular, the replacement processes, apparently, can also be evaluated through the methodology of P-parameter.

Methodology of Calculations

It is found out that during the self-diffusion the activation energy often equals the total of enthalpies of formation and transition of vacancies. Apparently, in any actual case, different diffusion mechanisms can perform, but activation energy has to be always defined by the specifics and degree of interatomic interactions in structures.

Thus, the task for estimating the activation energy, in many cases, comes to defining the actual energy of paired interatomic interaction of diffusion atom and atoms of diffusion medium – for each specific type of interaction.

Previously, diffusion coefficients of some refractory oxides were calculated via P-parameter of interacting atoms of diffusion and diffusion medium using initial spatial-energy notions [5].

Let us show that activation energy during the transition of atoms and molecules in different systems can be quite reliably and easily calculated with the help of the notion of P-parameter. For this we use the tabulated values of P_0-parameters calculated via the ionization energy by equations (5) – (8), some of them are given in Table 1. In diffusion processes, where ionic structures prevail, the ionisation energy of an atom (E_i) is used as the orbital energy, in contrast to purely covalence interactions, when the values of electron bond energy (W_i) are applied in calculations.

Let us define the averaged effective energy of paired interatomic interaction in the system M'- M" with the help of P-parameter. Summing up P_0 by valence electrons and dividing the value of $P_0/2r_i$ by the number of effective valence electrons of the atom (n), we obtain the atom affective energy falling on one valence electron: $Q = \dfrac{P_0}{2r_i n}$

Applying the previously established principle of adding inverse values of P_E-parameters – equation (10), we have the resultant value of the effective energy of paired interaction between atoms 1 and 2 in the process of diffusion and self-diffusion:

$$\frac{1}{E_a} = \frac{1}{Q_1} + \frac{1}{Q_2};\text{or} \quad \frac{1}{E_a} = 2\left[\left(\frac{r_i n}{P_0}\right)_1 + \left(\frac{r_i n}{P_0}\right)_2\right] \qquad (12)$$

where $E_a$is activation energy.

Here and afterwards the introduction of the multiplier 2 for r_i usuallycharacterises the volume diffusion, and without it – the surface one.

Table 1. *P*-parameters of valence orbitals of neutral atoms in basic state (calculated via the energy of atom ionisation)

Atom	Electrons of valence orbitals	E_i(eV)	r_i(Å)	q_i^2 (eVÅ)	P_0(eVÅ)	$\sum P_0$ (eVÅ)
1	2	3	4	5	6	7
H	$1s^1$	13.595	0.5295	14.394	4.7985	4.7985
Li	$2s^1$	5.390	1.586	5.890	3.487	3.487
C	$2p^1$	11.260	0.596	35.395	5.641	51.739
	$2p^1$	24.383	0.596	35.395	10.302	
	$2s^1$	47.86	0.620	37.243	16.515	
	$2s^1$	64.48	0.620	37.243	19.281	
O	$2p^1$	13.618	0.414	71.380	5.225	17.304
	$2p^1$	35.118	0.414	71.380	12.079	
Na	$3s^1$	5.138	1.713	10.058	4.694	4.694
Mg	$3s^1$	7.469	1.279	17.501	6.274	15.436
	$3s^1$	15.035	1.279	17.501	9.162	
P(III)	$3s^1$	10.487	0.919	38.199	7.696	35.996
	$3p^1$	19.73	0.916	38.199	12.268	
	$3p^1$	30.16	0.916	38.199	16.038	
K	$4s^1$	4.339	2.162	10.993	5.062	5.062
Ca	$4s^1$	6.113	1.690	17.406	6.483	15.803
	$4s^1$	11.871	1.690	17.406	9.320	
Cl	$3p^1$	12.268	0.728	59.842	8.125	8.125
N	$2p^1$	29.60	0.488	52.912	6.257	33.664
	$2p^1$	47.426	0.487	52.912	11.319	
	$2p^1$	77.472	0.487	52.912	16.078	
Mn(II)	$4s^1$	7.435	1.278	25.118	6.895	18.025
	$4s^1$	154640	1.278	25.118	11.130	
Mn(III)	$3d^1$	33.69	0.3885	177.33	12.200	30.225
Fe(II)	$4s^1$	7.893	1.227	26.57	7.098	18.462
	$4s^1$	16.183	1.227	26.57	11.364	
Fe(III)	$3d^1$	30.64	0.365	199.95	10.564	29.026
Fe(II)	$4s^1$	7.893	1.227	26.57	7.098	12.835
	$3d^1$	16.183	0.365	199.95	5.7372	
Fe(III)	$3d^1$	30.64	0.365	199.95	10.564	23.399
Tl	$6p^1$	6.108	1.319	60.054	7.037	7.037

The results of calculations E_a by equation (12) for many binary systems are given [6].

To calculate the activation energy of diffusion processes in binary systems, first, the effective energy of interatomic interaction is found, for example, for the pair metal-oxygen (in oxides) or metal (1)-metal (2) in metal systems, based on the following equations:

$$\frac{1}{Q_C} = \frac{1}{Q_i^{'}} + \frac{1}{Q_i^{''}} \quad (13)$$

$$Q_i^{'} = \left(\frac{P_0}{r_i n}\right)_1 ; \quad (13a)$$

$$Q_i^{''} = \left(\frac{P_0}{r_i n}\right)_2 \quad (13b)$$

where $Q_i^{'}$ and $Q_i^{''}$ are effective P-parameters of atoms 1 and 2, n_1 and n_2 – the number of their valence electrons in chemical bond. Since the activation energy of diffusion in this model corresponds to the total effective energy of interatomic interaction in the system: diffusing atom–diffusion medium, we have:

$$\frac{1}{E_a} = \frac{1}{Q} = \frac{1}{Q_1} + \frac{1}{Q_C} \quad (14)$$

where Q_1 is the value of effective P-parameter of the diffusing atom:

$$Q_1 = \frac{P_0}{r_i n} \quad (15)$$

Activation energy of diffusion in biosystems. Based on the theory of M. V. Volkenshtein, the transition of ions through the membrane can be explained as the motion of special quasi-particles ('conformones') consisting of the carriers of electron charge or shift of electron density. 'There is a reason to believe that conformation motion represents a limited continuous diffusion in extended potential pit' [11]. It means that the transport of atoms and molecules through biosystems is considered as the diffusion process defined by the changes in electron density during electron conformation interactions (ECI).

Spatial-energy parameter (P-parameter) is a direct characteristic of electron density of the atom at the given distance from the nucleus. Therefore, it is advisable to apply its methodology to evaluate the diffusion energy in biomembranes. The nature has chosen only several main elements participating in vital physical and chemical processes of biodiffusion. These are: sodium, potassium, chlorine, calcium, magnesium, hydrogen, phosphorus and their compounds. Transport can be passive, active and electrodiffusion.

Based on equation (12), we calculated the energy parameters of paired bond (Q_C) in the systems M-O (where M - Fe, Mg, Ca), P-O and M′-M″. Orbital radii (r_i) for metal systems and ionic radii (R_I) – in the chemical bond P-O were used as dimensional characteristics. The calculation results are given in Table 2. In the last column of Table 2 in the denominator (for metal systems) we see experimental values of activation energy E_a (Ref. 12). The direction of diffusion process is marked with arrow.

The values of activation energy in some oxides of biosystems were calculated using the obtained values of Q_C by equations (14) and (15). Here (as before) for metals, when calculating Q_1, orbital radii (r_i) were used as a dimensional characteristic, and for oxygen ions – anion dimensions (1.36 Å)(Table 3).

The experimental data [13, 14] for systems CaO and MgO are given for comparison. The activation energy of diffusion of ions Ca^{2+} via phosphor-lipids of membranes were evaluated (in the first approximation) as structural interactions of calcium ions with the bond P-O. Experimental values of activation energy, energy of ATP and its free energy (ΔG) – at different valence-structural states (conformations) of initial systems are given for comparison.

Table 2. Calculations of energy parameters of paired bond (Q_C)

Bond	First element					Second element					Q_i (eV)
	P_0 (eVÅ)	r_i (Å)	R_i (Å)	n	Q'_i (eV)	P_0 (eVÅ)	r_i (Å)	R_i(Å)	n	Q''_i (eV)	
Fe-O	29.026	1.227	($3d^1 4s^2$)	1	23.656	17.304	0.414		1	41.797	15.107
Mg-O	15.436	1.279	($3s^2$)	1	12.068	17.304	0.414		1	41.797	9.3646
Ca-O	15.803	1.690	($4s^2$)	1	9.3509	17.304	0.414		1	41.797	7.6415
Mg-O	15.436	1.279	($3s^2$)	2	6.0345	17.304	0.414		2	20.899	4.6825
P-O	7.696	($3P^1$)	2x1.86	5	0.4138	5.225	($2p^1$)	2×1.36	2	0.9605	0.2892
P-O	7.696	($3P^1$)	1.86	5	0.8275	5.225	($2p^1$)	1.36	2	1.9210	0.5784
P-O	7.796	($3P^1$)	1.86	3	1.3792	5.225	($2p^1$)	1.36	2	1.9210	0.8028
C→Ca	15.943	$2P^2$	2.60	2	3.066	15.803	($4s^2$)	1.04	2	7.5976	1.092/1.010
Fe→Ca	12.835	($4S^1 3d^1$)	2×0.80	2	4.011	15.803	($4s^2$)	1.04	2	7.5976	1.31/1.29
Mn→Fe	18.025	($4S^2$)	0.91	2	9.9038	29.026	($4s^2 3d^1$)	0.67	3	14.441	2.94/2.86
Mn→Fe	18.025	($4S^2$)	0.91	2	9.9038	18.462	($4s^2$)	0.80	2	11.539	2.67/2.71

Table 3. Activation energy of diffusion in oxides of biosystems

Diffusion element						Diffusion medium		E_a (eV)	
Atom	Orbitals	P_0(eVÅ)	r_i(Å) R_i	n	Q_1 (eV)	System	Q_C (eV)	calcu-lation	experiment
Ca	$4s^2$	15.803	1.690	2	4.676	CaO	7.6415	2.90	2.79
O^{-2}	$2p^2$	17.304	R_i=1.36	2	6.3618	CaO	7.6415	3.47	3.50
Mg	$3s^2$	15.436	1.279	2	6.0345	MgO	9.3646	3.67	3.46
O^{-2}	$2p^2$	17.304	R_i=1.36	2	6.3618	MgO	9.3646	3.79	3.56
O^{-2}	$2p^1$	5.225	R_i=1.36	1	3.8419	MgO	9.3646	2.72	2.71
O^{-2}	$2p^1$	5.225	R_i=1.36	2	1.9210	MgO	4.6825	1.36	1.31
Mg^{2+}	$4s^2$	18.025	2×0.91	7	1.4149	MgO	9.3646	1.23	1.21
Ca^{2+}	$4s^1$	6.483	$2R_i$=2.08	2	1.5584	P-O	0.2892	0.244 (0.203)	0.207
Ca^{2+}	$4s^1$	6.483	$2R_i$=2.08	2	1.5584	P-O	0.5784	0.4218 (0.405)	0.414 ATP hydrolysis
Ca^{2+}	$4s^1$	6.483	R_i=1.04	2	3.1168	P-O	0.8028	0.638	0.59-0.67 ΔG ATP

Table 4. Diffusion via biomembrane with movable carriers (MC)

Diffusing atom	Orbital	P_0 (eVÅ)	r_i(Å)	n	$P'_E = \frac{P_0}{r_i n}$ (eV)	Carriers	$E_a \approx$ (eV)	α (%)	ρ (%)	Transport type
Na	$3s^1$	4.694	1.713	1	2.740	valinomycine	2.393	13.5	15-18	MC
Na	$3s^1$	4.694	2×1.713	1	1.370	monactine	1.414	3.16	100	MC
K	$4s^1$	5.062	2.162	1	2.341	valinomycine	2.393	2.20	100	MC
Cl	$3p^1$	8.125	0.728	7	1.594	monactine	1.414	12.0	17-22	MC
Cl	$3p^1$	8.125	0.728	5	2.232	valinomycine	2.393	6.96	92-96	MC
Ca	$4s^2$	15.803	1.690	4	2.238	valinomycine	2.393	6.69	93-97	MC
Mg	$3s^2$	15.436	1.279	4	3.017	valinomycine	2.393	23.1	0-5	no MC
Na	$3s^1$	4.694	1.713	1	2.740	gramicidin	0.404	149	0	no MC
K	$4s^1$	5.062	2.162	1	2.3414	gramicidin	0.404	141	0	no MC
Cl	$3p^1$	8.125	0.728	7	1.594	gramicidin	0.404	119	0	no MC

Table 5. Diffusion through the pores with gramicidin

Diffusing elements							Pores	α (%)	ρ (%)
Atom	Orbitals	P_0 (eVÅ)	r_i (Å)	n_1-n_2	Q_1(eV) by eq. (12)	Process	ΔH (eV)		
Na	$3S^1$	4.694	1.713	1-2	0.457	self-diffusion	0.404	12.3	17-22
K	$4S^1$	5.062	2.162	1-2	0.390	self-diffusion	0.404	3.53	100
Cl	$3P^1$	8.125	0.728	7-7	0.399	self-diffusion	0.404	1.25	100

From Tables 2 and 3 we see that calculated values mainly correspond to the data in the limits of allowed errors of calculations and experiment.

It is known [15] that the transition of Ca^{2+} from the one side of the membrane onto the other is connected with the consumption of energy 0.207 eV, and that the transition of two ions of calcium is connected with the hydrolysis of one ATP molecule. FromTable 3 we have, for two ions of Ca: $2E_a$=2 ×0.244=0.488 eV that somewhat exceeds the ATP hydrolysis energy that equals 0.414 eV. However, if we consider that the efficiency coefficient for actual processes during photosynthesis is 83%, for two ions of calcium we have: $2E_a$=0.488 ×0.83=0.405 eV (and E_a≈0.203 eV – for one ion of calcium), this is in a good accordance with the reference data.

To evaluate the energy of diffusion transport, the values of P_E-parameters of diffusing atoms Na, K, Cl and Mg were calculated with the help of movable carriers (MC) (Table 4). The values of activation enthalpy (ΔH) of ionophores were taken from Ref. 11. At the same time, it was considered that these values can be approximately (if the activation barrier equals the enthalpy of the process [16]) taken as equal to the activation energy (E_a):

$$E_a \approx \Delta H$$

According to the initial principles of this methodology or the nomogram (Figure 1), the complete (ρ=100%) structural interaction takes place at α<6%, limited interaction – at 6%<α<(25-30)%. When $P''_E = \Delta H$, the coefficient α was calculated byequation (11) and (by nomogram) – the degree of structural interaction of diffusing atoms (ρ) with ionophores valinomycine and monaktine. In Table 4 one can see the results of calculations and principal possibility of the process, its degree or absence.

K, Ca and Cl have total or almost total structural interaction with valinomycine, and Na – with monaktine. Na has limited conformation interaction with valinomycine, and Cl – with monaktine.The lack of the possibility of Mg to transit with these ionoforms shows, in particular, why the nature chose calcium, but not magnesium as the main regulator of various intracellular processes. From Table 4 we see the complete absence of diffusion with the movable carrier of these elements with gramicidin (α>100%), but gramicidin forms a polar pore in biomembrane. The ion is carried not by one, but several molecules of ionophore. This is a forwarding mechanism where the ions of diffusing element, apparently, move by the principle of self-diffusion in energy field of gramicidin molecules. In Table 5 one can see the calculations of activation energy (E_a=Q_1) of self-diffusion of sodium, potassium and chlorine ions – byequation (12), where the possibility of generalising two-valence electrons of the medium atom for Na and K is taken into account (n_1=1 and n_2=2).

There one can also find the results of calculating the coefficient α and defining the degree of structural interaction (ρ) during the diffusion of ions Na, K and Cl through the pores with gramicidin. It is shown that ion Na^+ has a limited diffusion, and ions K^+ and Cl^- – total. This takes place under standard conditions. Conformation changes caused by external effects (for example, temperature) can produce certain shifts in optimal values of these criteria.

Relative Penetrability of Biomembrane Channels

In literature [3,4] much attention is paid to geometrical criteria of penetrability. Obviously, these criteria are necessary factors of penetrability, but, apparently, insufficient, and, perhaps, even not the main ones. Thus, we can see based on hydroxyl-ammonium and methyl-ammonium[4] that these structures, having practically the same dimensions, possess completely different penetrability via the biochannels of sodium and potassium.

Using the methodology of *P*-parameter, we can assume that the availability of direct functional dependence between the value of relative penetrability of membranes and degree of structural interaction (ρ):

$$\gamma \sim \rho \tag{16}$$

where ρ is determined by the nomogram based on the calculated values of relative difference of *P*-parameters of coefficient (α) of interacting structures – equation (11).

This can be easily demonstrated on the example of Na- and K-channels for 1-valence ions Li^+ and Tl^+ (Tables 6 and 7).

Here, *P*-parameters are calculated by equation (6), in which the ion radiusR_I is used as a dimensional characteristic, and the coefficient α is calculated by equation (11). The value of ρ (which is not expressed in %) is defined by the nomogram (Figure 1).

Table 6. StructuralP_S-parameters of hydroxyl-ammonium and methyl-ammonium

Mole-cule complex	1st component			2nd component			Complex	Calculation formulas
	P_0 (eVÅ)	R_i (Å)	P_E (eV)	P_0 (eVÅ)	R_i (Å)	P_E (eV)	n_c(eV)	
NH_3^+	33.664	1.48	22.746	3×4.7985	1.36	10.575	7.2190	$\frac{1}{P'_E}+\frac{1}{P''_E}=\frac{1}{P'_S}$
OH	5.225	1.36	3.8419	4.7985	1.36	3.525	1.8383	$\frac{1}{P'_E}+\frac{1}{P''_E}=\frac{1}{P''_S}$
NH_3^+-OH							9.057	$P_S=P'_S+P''_S$
CH_3	51.739	2.60	19.900	3×4.7985	1.36	10.575	6.9066	$\frac{1}{P'_E}+\frac{1}{P''_E}=\frac{1}{P''_S}$
NH_3^+-CH_3							14.125	$P_S=P'_S+P''_S$

Table 7. Calculation of relative penetrability in the channels of biomembranes

Channels atoms	Main atom of the channel			Diffusion atom or complex			α (%)	ρ (not in %)	γ (Ref. 14)
	P_0 (eVÅ)	R_i (Å)	P_E (eV)	P_0 (eVÅ)	R_i (Å)	P_E (eV)			
Na-Li	4.694	0.98	4.790	3.487	0.68	5.128	6.814	0.9-0.95	0.9
K-Li	5.062	1.33	3.806	3.487	0.68	5.128	29.60	below 0.01	0.02
Na-Tl	4.694	0.98	4.790	7.037	1.36	5.223	8.654	0.35-0.4	0.33
K-Tl	5.062	1.33	3.806	7.037	1.36	5.223	31.40	below 0.01	?
2Na-(H_3N-OH)	2×4.694	0.98	9.580			9.057	5.61	~1.0	~(1-0.9)
2Na-(H_3N-CH_3)	2×4.694	0.98	9.580			14.125	38.3	~0	0.007
2Na-(H_3N-CH_3)	2×5.062	1.33	7.612			14.125	59.9	~0	0.02

For complicated complex structures it is necessary to determine their total structural P_S-parameter, using the above rules of adding *P*-parameters. In each separate fragment of the molecule P_S-parameter is calculated by the principle of adding the inverse values of initial parameters, and *P*-parameters of such fragments are added algebraically. In Table 6 are presented these formulas and results of calculating *P*-parameters for hydroxyl-ammonium and methyl-ammonium, and in Table 7 – the obtained values of α and ρ in these cases.

It should be noted that less accurate values of ρ are obtained by the nomogram for the values α>(25-30)%. In all other cases, the data obtained are rather close to the actual values of ρ and γ.

Conclusions

The method for calculating the activation energy of diffusion processes in biomembranes isbased on the initial spatial-energy characteristics of free atoms.

Relative penetrability of biomembranes numerically equals the degree of structural interaction calculated by the nomogram via the relative difference of P-parameters of interacting structures.

The calculated values are mainly in accordance with experimental and reference data (within the limits of possible errors of experiments and calculations).

References

[1] Z. P. Kometiani, M. G. Vekua: Kinetics of Membrane Transport Ferments . In: *Biochemistry of Membranes.* Vol. 5, Vysshaya Shkola, Moscow, 1988, 111 p.

[2] A. N. Remizov: *Medical and Biological Physics.* Vysshaya Shkola, Moscow, 1987, 638 p.

[3] A. B. Rubin: Biophysics. Book 1. *Theoretical Biophysics.* Vysshaya Shkola, Moscow, 1987,319 p.

[4] A. B. Rubin: Biophysics. Book 2. Biophysics *of Cellular Processes.* Vysshaya Shkola, Moscow, 1987, 303 p.

[5] G. A. Korablev: Calculation of Diffusion Coefficient of Refractory Oxides of Metals Based on the Use of Spatial-energy Parameters of Neutral Atoms. In: Diffusion, Sorption and Phase Transformations in the Processes of Metal Reduction. *Nauka,* Moscow, 1981, 4-6.

[6] G. A. Korablev: Spatial – energy Principles of Complex Structures Formation. *Brill Academia Publishers and VSP,* Leiden, The Netherlands, 2005, 426 p. (monograph).

[7] C. F. Fischer: Average-Energy of Configuration Hartree-Fock Results for the Atoms Helium to Radon.*Atomic Data,* (4),301 (1972).

[8] K. U. Alen: *Astrophysical Values. Mir,* Moscow, 1977, 446 p.

[9] J. T. Waber, D. T. Cromer: Orbital Radii of Atoms and Ions. *J. Chem. Phys.,*42 (12), 4116 (1965).

[10] E. Clementi, D. L. Raimondi: Atomic Screening Constants from S.C.F. Functions, 1. *J. Chem. Phys.,*38 (11), 2686 (1963).

[11] M. V. Volkenshtein: Biophysics. *Nauka,* Moscow, 1988, 592p.

[12] B. S. Bokshtein: *Diffusion in Metals. Metallurgia,* Moscow, 1978, 248 p.

[13] G. V. Samsonov (Ed.): *Physical and Chemical Properties of Oxides.* Metallurgia, Moscow, 1978, 472 p.

[14] P. Kofstad: Deviation from Stoichiometry, Diffusion and Electrical Conductivity in Simple Oxides of Metals. *Mir,* Moscow, 1975, 398 p.

[15] D. O. Levitskiy: Calcium and Biological Membranes. In: *Biochemistry of Membranes.* Vol. 7, Vysshaya Shkola, Moscow, 1990, 126 p.

[16] L. A. Blyumenfeld: *Problems of Biological Physics*. Nauka, Moscow, 1977, 336 p.

In: Chemical and Biochemical Kinetics: New Perspectives ISBN 978-1-61209-224-9
Editor: Gennady Zaikov

Chapter 2

MATHEMATICAL MODELLING DEPENDENCE OF THERMODYNAMIC CHARACTERISTICS UPON SPATIAL-ENERGY PARAMETER OF FREE ATOMS

G. A. Korablev[1,*], N. G. Korableva* and G. E. Zaikov[2†]
[1]Izhevsk State Agricultural Academy, Russia, Izhevsk, Russia
[2]N.M. Emanuel Institute of Biochemical Physics, RAS, Moscow, Russia

ABSTRACT

The dependence of some thermodynamic characteristics upon initial spatial-energy parameters of free atoms has been analyzed. The corresponding equations have been obtained, the dissociation energies of binary molecules and enthalpy of single-atom gas formation have been calculated based on them.

Keywords: Spatial-energy parameter, formation enthalpy, sublimation, dissociation energy

INTRODUCTION

Thermodynamic parameters (enthalpy, entropy, thermodynamic potential) allows describing various physical, chemical and other processes and explicitly assess the possibility of their flow without vividly using physical models. The application of reliable values of formation enthalpies is required for searching new perspective materials and compounds, assessment of molecule kinetic properties, analysis mechanisms of rocket propellant combustion, etc. [1,2,3].

The establishment of connection between the structure and thermochemical parameters, as well as kinetic and thermochemical characteristics of interacting systems is of great

[*] E-mail: biakaa@mail.ru, korablev@udm.net; Tel.: +7(3412) 591946.
[†] E-mail: chembio@sky.chph.ras.ru.

importance [4,5]. But the analysis of dependences between main parameters of chemical thermodynamics and spatial-energy characteristics of free atoms is still topical.

For this purpose the methodology of spatial-energy parameter (P-parameter) has been used in this research [6].

RESEARCH TECHNIQUE

The comparison of multiple regularities of physical and chemical processes allows assuming that in many cases the principle of adding reciprocals of volume energies or kinetic parameters of interacting structures is fulfilled.

Some examples are: ambipolar diffusion, cumulative rate of topochemical reaction, change in the light velocity when moving from vacuum into the given medium, resultant constant of chemical reaction rate (initial product – intermediary activated complex – final product).

Lagrangian equation for the relative motion of isolated system of two interacting material points with masses $m_1$and$m_2$in the coordinate x with acceleration α can be as follows:

$$\frac{1}{1/(m_1 a\Delta x)+1/(m_2 a\Delta x)} \approx -\Delta U \quad \text{or} \quad \frac{1}{\Delta U} \approx \frac{1}{\Delta U1} + \frac{1}{\Delta U2} \tag{1}$$

where ΔU_1 and ΔU_2 – potential energies of material points on elemental interaction section, ΔU – resultant (mutual) potential energy of these interactions.

The atom system is formed by oppositely charged masses of nucleus and electrons. In this system the energy characteristics of the subsystems are: orbital energy of electrons and effective energy of nucleus considering screening effects (by Clementi). Either the bond energy of electrons (W) or the ionization energy of an atom (E_i) can be used as the orbital energy.

Therefore, assuming that the resultant interaction energy in the system orbital-nucleus (responsible for interatomic interactions) can be calculated following the principle of adding reciprocals of some initial energy components, the introduction of P-parameter [6,7] as averaged energy characteristic of valent orbitals based on the following equations has been substantiated:

$$\frac{1}{q^2/r_i} + \frac{1}{W_i n_i} = \frac{1}{P_э} \tag{2}$$

$$P_E = \frac{P_0}{r_i} \tag{3}$$

$$\frac{1}{P_0} = \frac{1}{q^2} + \frac{1}{(wrn)_i} \tag{4}$$

$$q = \frac{Z^*}{n^*} \tag{5}$$

where: W_i – bond energy of an electron [8]; n_i – number of elements of the given orbital; r_i– orbital radius of *i*–orbital [9]; Z* and n* – effective charge of a nucleus and effective main quantum number [10,11].

The value of P_0 will be called spatial-energy parameter, and the value of P_E – effective P-parameter. The effective P_E-parameter has a physical sense of some averaged energy of valent electrons inside the atom and is measured in energy units, e.g. in electron-volts (eV).

The calculations demonstrated that the values of P_E-parameters numerically equal (in the limits of 2%) the total energy of valent electrons (U) by statistic model of an atom. Using the known relationship between the electron density (β) and interatomic potential by statistic model of an atom, the direct dependence of P_E-parameter upon the electron density on the distance $r_i$from the nucleus can be obtained:

$$\beta_i^{2/3} = \frac{AP_0}{r_i} = AP_E$$

where A – constant.

Based on the equations (2-5) P_E andP_0-parameters of free atoms for the majority of elements of the periodic system have been calculated [6,7]. Some of these calculations are given in Table 1.

Modifying the rules of addition for reciprocals of energy values of subsystems with reference to complex structures, the formula for calculating P_C-parameter of a complex structure is obtained:

$$\frac{1}{P_C} = \left(\frac{1}{NP_E}\right)_1 + \left(\frac{1}{NP_E}\right)_2 + \ldots \tag{6}$$

where N_1 and N_2 – number of homogeneous atoms in subsystems.

Calculation of Dissociation Energy of Binary Molecules via Average Values of P_o-Parameters

The application of methods of valent bond and molecular orbitals to complex structures faces significant obstacles for prognosticating energy characteristics of the bonds formed.

Based on the equation (6), the application of the formula for calculating the dissociation energy has turned out to be practically suitable [12]:

$$\frac{1}{D_0} = \frac{1}{P_C} = \frac{1}{\left(P_E \frac{N}{K}\right)_1} + \frac{1}{\left(P_E \frac{N}{K}\right)_2} \quad (7),$$

where N – bond order,K – maxing or hybridization coefficient that usually equals the number of valent electrons considered, and the value of P_E (N/K) has a physical sense of the averaged energy of spatial-energy parameter accrued to one valent electron of the orbitals registered. For complex structures P_E-parameter was being averaged by all main valent orbitals.

For binary molecules the dissociation energy (D_o) corresponds to the value of chemical bond energy: $D_o=E$.

The calculation results of dissociation energy by the equation (7), given ion Table 2, demonstrated that $P_C=Д_0$. For some molecules containing such elements as F, N and O, the values of ion radius have been applied to register the bond ionic character for calculating P_E-parameter (in Table 2 marked with *). For such molecules as C_2, N_2, O_2 the calculations have been made by multiple bonds. In other cases the average values of bond energy have been applied. The calculated data do not contradict the experimental ones [2, 3].

Equation of Dependence of Thermodynamic Characteristics upon Spatial-Energy Parameters of Free Atoms

The dissociation energy (E) of the molecule breakage into two parts numerically equals the difference of the heat produced during the formation of dissociation products and during the initial molecule formation:

$$E = D_0 = \left[\Delta H_0(R_1) + \Delta H_0(R_2)\right] - \Delta H_0(R_1 R_2) \quad (8)$$

where D_0 – dissociation energy of a molecule (rupture energy of its bond), $\Delta H_0(R_1)$ and $\Delta H_0(R_1)$ – formation enthalpy at 0°K respectively of dissociation products R_1 and R_2, $\Delta H_0(R_1 R_1)$ – initial energy formation enthalpy.

Since for binary molecules$E=D_0$, the equation (7) gives:

$$E = \frac{\left(P_E \frac{N}{K}\right)_1 \left(P_E \frac{N}{K}\right)_2}{\left(P_E \frac{N}{K}\right)_1 + \left(P_E \frac{N}{K}\right)_2} \quad (7a)$$

From the formulas (8) and (7a) we have:

$$\frac{\left(P_E\frac{N}{K}\right)_1\left(P_E\frac{N}{K}\right)_2}{\left(P_E\frac{N}{K}\right)_1+\left(P_E\frac{N}{K}\right)_2}=\Delta H_0(R_1)+\Delta H_0(R_2)-\Delta H_0(R_1R_2) \qquad (9)$$

This is the equation of direct dependence of thermodynamic values and initial spatial-energy characteristics of a free atom.

The relationship obtained (9) allows defining the corresponding dependences with the experimental thermodynamic characteristics of chemical reactions.

Thus in thermodynamic experimental methods for determining the formation enthalpy the following equation is used:

$$\Delta H_0 = T\left(\Delta \Phi_T^* - R\ln K_E\right)$$

where: K_E – equilibrium constants of chemical reaction; T – thermodynamic temperature of the process; $\Delta\Phi_T^*$ – change in the reaction of considered reduced thermodynamic potential; R – universal gas constant.

EVALUATION OF SINGLE-ATOM GAS FORMATION ENTHALPY (ΔH_G^0)

In the experimental evaluation method of single-atom gas formation enthalpy from solids, a known thermodynamic identity can be used:

$$\Delta H_G^0 \equiv \Delta H^0_{SD} + \Delta H_{S,298} \qquad (10)$$

where: ΔH_G^0 – formation enthalpy of gaseous substance;
ΔH^0_{SD} – formation enthalpy of solid in nonstandard state;
$\Delta H_{S,298}$ – sublimation enthalpy (substance transition from solid into gaseous state).

Table 1. P-parameters of atoms calculated via bond energy of electrons

Atom	Valent electrons	W (eV)	r_i (Å)	$q^2{}_0$ (eVÅ)	P_0 (eVÅ)	R (Å)	P_0/R (eV)	$P_0/$ $R(n^*+1)$	r_i (Å)	P_0/r_i (eV)	$P_0/$ $r_i(n^*+1)$
1	2	3	4	5	6	7	8	9	10	11	12
Li	$2S^1$	5.3416	1.506	5.86902	3.475	1.55	2.2419	0.7473	0.68	5.1103	1.7034
Na	3S1	4.9552	1.713	10.058	4.6034	1.69	2.4357	0.60892	0.96	4.6973	1.1743
K	4S1	4.0130	2.612	10.993	4.8490	2.36	2.0547	0.4372	1.33	3.6459	0.7757
Rb	5S1	3.7511	2.287	14.309	5.3630	2.48	2.1625	0.43250	1.49	3.5993	0.71986
Cs	6S1	3.3647	2.516	16.193	5.5628	2.68	2.0757	0.3992	1.65	3.3714	0.6483
Mg	3S1	6.8859	1.279	17.501	5.8568	1.60 1.60	3.6616	0.91544	0.74	7.9173	1.9793
	3S2	6.8859	1.279	17.501	8.7787		514667	1.3717	0.74	11.863	2.9658
Ca	4S1	5.3212	1.690	17.406	5.929	1.97 1.97	3.0096	0.64035	1.33		
	4S2	5.3212	1.690	17.406	6.6456		4.902	0.95535	1.04	8.5054	1.8097
Sr	5S1	4.8559	1.836	21.224	6.790	2.15 2.15	2.9205	0.56409			
	5S2				9.6901		4.5070	0.90140	1.20	8.0751	1.6150
Ba	6S1	4.2872	2.060	22.950	6.3768	2.21 2.21	2.6854	0.5549	1.38	7.2328	1.3909
	6S2				9.9812		4.5164	0.8685			
	4S1	5.7174	1.570	19.311	6.I279	1.64	3.7365	0.7950			
Sc	4S2	5.7174	1.570	19.311	9.3035						
	3d1	9.3532	0.539	81.099	4.7463	1.64 1.64	2.8941	0.6158			
	4S2+3d1	9.3532	0.539	81.099	14.050		8.5671	1.8228	0.83	16.928	3.6016
	5S1	6.3376	1.693	22.540	6.4505	1.81	3.5638	0.71276			
Y	5S2				10.030		5.5417	1.1083			
	4d1	6.7965	0.856	229.18	5.6756		3.1357	0.62714			
	4d1+5S2				15.706		8.6771	1.7354	0.97	16.192	3.2384
	6S1	4.3528	1.915	34.681	6.7203	1.87	3.5937	0.6911			
	6S2				11.259		6.0209	1.1579			
	4f1	10.302	0.4234	17870	4.2576		2.2768	0.4378			
La	5P1	25.470	0.827	145.53	18.400		9.8396	1.8922			
	6S2+5P1				29.659		15.860	3.0501	1.04	28.518	5.4843
	6S2+4f1				15.517		8.2976	1.5957	1.04	14.920	2.8693

Atom	Valent electrons	W (eV)	ri (Å)	q2 0 (eVÅ)	P0 (eVÅ)	R (Å)	P0/R (eV)	P0/ R(n*+1)	ri (Å)	P0/ri (eV)	P0/ ri(n*+1)
Ti	4S1	6.0082	1.477	20.879	6.2273	1.46 1.46	4.2653	0.9075			
	4S2	6.0082	1.477	20.879	9.5934	1.46	6.5708	1.3960	0.78	12.299	2.6169
	3d1	11.990	0.496	106.04	5.556		3.8053	0.80965			
	3d2	11.990	0.469	106.04	10.558	1.46					
	4S2+3d2				20.151		13.802	2.9366	0.64	31.486	6.6981
	4S1+3d1							1.7172			
Zr	5S1	5.5414	1.593	23.926	6.5330	1.60	4.0831	0.8166			
	5S2				10.263		6.4146	1.2829			
	4d1	9.1611	0.790	153.76	6.9121		4.3201	0.8640			
	4d2				13.229		8.2681	1.6536			
	5S2+4d2				23.492	1.6	14.683	2.9365	0.82	28.640	5.7298

Table 2. Dissociation energies of two-atom molecules – $D_0\left(\frac{KJ}{mol}\right)$

Structure	First atom				Second atom				P_C (eV)	D_0 calcul.	D_0 experim.
	Orbitals	N/k	P_E (eV)	$P_E\frac{N}{k}$	Orbitals	N/k	P_E (eV)	$P_E\frac{N}{k}$			
1	2	3	4	5	6	7	8	9	10	11	12
C*Cl*	$2P^1$	1/1	7.6208	7.6208	$3P^1$	1/1	8.5461	8.5461	4.0209	388.9	393.3
CBr	$2P^1$	1/1	7.6208	7.6208	$4P^1$	1/1	8.0430	8.0430	3.9130	377.7	364
CJ	$2P^1$	1/1	7.6208	7.6208	$5P^1$	1/1	7.2545	7.2545	2.2523	217.4	209.2
CN	$2P^2$	2/2	13.066	13.066	$2S^22P^3$	2/5	47.413	18.965	7.7358	746.7	755.6
CN	$2P^2$	2/2	14.581	14.581	$2P^3$	2/3	25.127	16.751	7.796	752.5	755.6
C-O	$2P^2$	1/2	13.066	6.533	$2P^2$	1/2	17.967	8.984	3.782	365	356
NO	$2P^1$	1/1	9.2839	9.2839	$2P^2$	2/2	20.048	20.048	6.346	612.5	626.8
CH	$2P^2$	1/2	13.066	6.533	$1S^1$	1/1	9.066	9.066	3.7969	366.5	333±1
OH	$2P^2$	1/2	17.967	8.9835	$1S^1$	1/1	9.066	9.066	4.5118	435.5	423.7

Table 2. Dissociation energies of two-atom molecules – $D_0\left(\frac{KJ}{mol}\right)$ (Continued)

Structure	First atom				Second atom				P_C (eV)	D_0 calcul.	D_0 experim.
	Orbitals	N/k	P_E (eV)	$P_E\frac{N}{k}$	Orbitals	N/k	P_E (eV)	$P_E\frac{N}{k}$			
C*l*F	$3S^23P^5$	1/7	29.391*	4.1987	$2S^22P^5$	1/7	38.202*	5.4574	2.5579	246.9	229.1
C*l*O	$3S^23P^5$	1/7	29.391*	4.1987	$2P^2$	2/2	8.7191*	8.7191	2.8337	273.5	264
C*l*O	$3P^1$	1/1	4.7216*	4.7216	$2S^22P^4$	1/6	30.738*	5.123	2.450	237.2	264
FO	$2P^1$	1/1	4.9887*	4.9887	$2P^2$	1/2	8.7191*	4.3596	2.327	224.6	219.2
NF	$2P^3$	1/3	10.696*	3.5653	$2P^1$	1/7	38.202*	5.4574	2.486	239.5	298.9
NC*l*	$2P^3$	1/3	22.296	7.432	$3P^1$	1/1	8.5461	8.5461	3.9751	383.7	384.9
H_2	$1S^1$	1/1	9.0624	9.0624	$1S^1$	1/1	9.066	9.066	4.533	437.5	432.2
Li_2	$2S^1$	1/1	2.2419	2.2419	$2S^1$	1/1	2.2419	2.2419	1.121	108.2	98.99
B_2	$2P^1$	1/1	5.4885	5.4885	$2P^1$	1/1	5.4885	5.4885	2.744	264.9	276±21
C-C	$2P^1$	1/1	7.6208	7.6208	$2P^1$	1/1	7.6208	7.6208	3.810	367.8	376.7
C=C	$2P^2$	2/2	13.066	13.066	$2P^2$	2/2	13.066	13.066	6.533	630.6	611
N-N	$2P^3$	1/3	10.696*	3.5653	$2P^3$	1/3	10.696*	3.5653	1.783	172.1	161
N=N	$2S^22P^3$	2/5	22.745*	9.098	$2S^22P^3$	2/5	22.745*	9.098	4.549	439	418
O-O	$2P^2$	1/2	8.7191	4.3596	$2P^2$	1/2	8.7191	4.3596	2.1798	210.4	213.4
O=O	$2S^22P^4$	2/6	30.738*	10.246	$2S^22P^4$	2/6	30.738*	10.246	5.123	494.5	498.3

Table 3. Effective main quantum number

n	1	2	3	4	5	6
n^*	1	2	3	3.7	4	4.2
n^*+1	2	3	4	4.7	5	5.2

Table 4. calculation of formation enthalpy of single-atom gases – ΔH_{298}

1 eV = 96.525 kJ/mol

Group	Atom	Orbitals	P_o (eVA^o)	γ	$r(A^o)$	$\Delta H = \frac{P_o \gamma}{r(n^*+1)}$, (eV)	$r_u (A^o)$	$\Delta H = \frac{P_o \gamma}{r_u(n^*+1)}$, (eV)	ΔH calculation (kJ/mol)	ΔH° 298 ref. data (kJ/mol)
1	2	3	4	5	6	7	8	9	10	11
1a	Li	$2S^1$	3.4750	1			0.68	1.703	164.4	159.3
	Na	$3S^1$	4.6034	1			0.98	1.1743	113.3	107.5
	K	$4S^1$	4.8490	2	2.36	0.874			84.36	88.9
	Rb	$5S^1$	5.3630	2	2.48	0.865			83.49	80.9
	Cs	$6S^1$	5.5628	2	2.68	0.7984			77.07	77
2a	Be	$2S^2$	7.512	½			0.34	3.682	355.4	326.4
	Mg	$3S^2$	8.7787	½			0.74	1.483	143.1	147.1
	Ca	$4S^2$	8.8456	1			1.04	1.8097	174.7	177.8
	Sr	$5S^2$	9.6901	1			1.20	1.6150	155.9	160.7
	Ba	$6S^2$	9.9812	2	2.21	1.737			167.7	179.1
3a	Sc	$4S^2 3d^1$	14.050	1			0.74	4.040	389.96	379.1
	Y	$5S^1 4d^2$	17.527	2	1.62	4.342			419.2	423
	La	$6S^2 5p^1$ $6S^2 4f^1$	29.659 15.517	2 2	1.87 1.87	6.1002 3.1915			<448.4>	429.7

Table 4. calculation of formation enthalpy of single-atom gases – ΔH_{298}

$1\ eV = 96.525\ kJ/mol$ (Continued)

Group	Atom	Orbitals	P_o (eVA^o)	γ	$r(A^o)$	$\Delta H = \frac{P_o\gamma}{r(n^*+1)}$, (eV)	$r_u (A^o)$	$\Delta H = \frac{P_o\gamma}{r_u(n^*+1)}$, (eV)	ΔH calculation (kJ/mol)	$\Delta H^\circ 298$ ref. data (kJ/mol)
4a	Ti	$4S^1 3d^1$ $4S^2 3d^2$	11.7853 20.151	2 2	1.46 1.46	3.4344 5.8732			<449.2>	468.6
	Ti	$4S^1 3d^{1^1}$ $4S^2 3d^2$	11.7853 20.151	1 1			0.78 0.64	3.214 6.6981	<478.4>	468.6
	Zr	$5S^2 4d^2$	23.492	2	1.60	5.873			566.9	600
	Hf	$6S^2 5d^2$	24.498	2	$r_K = 1.44$	6.543			631.6	620.1
5a	V	$4S^2 3d^1$ $4S^1 3d^2$	15.776 17.665	1 1			0.67 0.67	5.6010 5.6090	<512.5>	514.6
	Nb ($5S^2 4d^3$)	$5S^2 4d^2$	25.577	1			0.67	7.6349	736.9	722.6
	Nb ($5S^1 4d^4$)	$5S^1 4d^2$ $5S^1 4d^4$	20.805 30.607	1 1			0.767 0.65	5.425 9.2748	<709.4>	722.6
	Nb (3, 4, 5)			1					<723.2>	722.6
	Ta	$6S^2 5d^2$ $6S^2 5d^3$	26.314 32.722	1 1			0.737 (0.66)	6.8662 9.5344	<791.5>	786.1

Group	Atom	Orbitals	P^o (eVA^o)	γ	$r\left(A^o\right)$	$\Delta H = \frac{P_o\gamma}{r(n^*+1)}$, (eV)	$r^u\left(A^o\right)$	$\Delta H = \frac{P_o\gamma}{r_u(n^*+1)}$, (eV)	ΔH calculation $\left(kJ/mol\right)$	ΔH° 298 ref. data $\left(kJ/mol\right)$
6a	Cr	$4S^2$ $4S^2\,3d^1$	10.535 17.168	1 1			0.83 0.64	2.7006 5.7074	<406.1>	397.5
	Mo (2) ($5S^2\,4d^4$)	$5S^1\,4d^1$	19.574	1			0.737	5.312	<658>	656.5
	Mo (4) ($5S^1\,4d^4$)	$5S^1\,4d^3$	28.293	1			0.68	8.3215		
	W(4) W(5)	$6S^2\,6S^1$ $6S^2\,5d^3$	27.879 34.828	2	1.40 1.40	3.8295 4.7841			<831.5>	856.9
7a	Mn (2)	$4S^1\,3d^1$	12.924	1			0.91	3.0217	291.7	284.5
	Tc (3)	$5S^1\,4d^2$	23.866	2	1.36	7.0194			674.1	657
	Re (4)	$6S^2\,5d^2$	29.806	1			0.72	7.961	768.4	775.7
	Re (6)	$6S^2\,5d^4$	44.519	½			0.52	8.232	794.6	
	Re (4,6)								<781>	
8	Fe (2) Fe (3)	$4S^1\,3d^1$ $4S^2\,3d^1$	12.717 16.664	1 1			0.80 0.67	3.3822 5.2918	<418.6>	417.1
	Co (2) Co (3)	$4S^1\,3d^1$ $4S^2\,3d^1$	12.707 16.680	1 1			0.78 0.64	3.4662 5.5785	<436.5>	428.4
	Ni (2) Ni (3)	$4S^1\,3d^1$ $4S^1\,3d^2$	12.705 16.897	1 1			0.74 0.60	3.6530 5.992	<465.5>	429.3
	Ni (2)	$4S^1\,3d^1$	12.705	2	1.24	4.360			420.8	429.3

Table 4. calculation of formation enthalpy of single-atom gases – Δ H_{298}

1 eV = 96.525 kJ/mol (Continued)

Group	Atom	Orbitals	P_o (eVA^o)	γ	$r(A^o)$	Δ H = $\frac{P_o\gamma}{r(n^*+1)}$, (eV)	$r_u(A^o)$	Δ H = $\frac{P_o\gamma}{r_u(n^*+1)}$, (eV)	Δ H calculation (kJ/mol)	Δ H° 298 ref. data (kJ/mol)
	Ru (3)	$5S^1 4d^2$	23.636	1			0.68	6.982	671	656.8
	Rh (3)	$5S^2 4d^1$	21.114	1			0.75	5.6304	543.5	556.5
	Pd (2)	$5S^2$	12.057	1			0.64	3.7681	363.7	372.3
8	Ir (2) Ir (4)	$6S^1 5d^1$ $6S^2 5d^2$	17.631 30.790	2 2	1.35 1.35	5.023 8.772			<665.8>	669.5
	Pt (2)	$6S^1 5d^1$	17.381	1			0.6	5.971	576.4	565.7
1б	Cu (2)	$4S^1 3d^1$	13.242	1			0.80	3.5218	339.9	337.6
	Ag Ag (1)	$4d^1$ $4d^1$	9.7843 9.7843	2 2	r_k = 1.25 1.44	3.131 2.718			<282.3>	284.9
	Au (3)	$6S^1 5d^2$	27.536	1	1.44	3.6774			355.0	368.8
2б	Zn (1) Zn (2)	$3d^1$ $4S^2$	6.1153 11.085	1 1	1.39 1.39	0.9361 1.6968			<127.1>	130.5
	Cd (2)	$5S^2$	11.839	½			0.99	1.1959	115.4	111.8
	Cd (2)	$5S^1 4d^1$	17.145	½	1.56	1.0991			106.1	111.8
	Cd (1, 2)								<110.8>	111.8
	Hg (1)	$5d^1$	11.266	½	1.60	0.6771			65.35	61.40

Group	Atom	Orbitals	P^o (eVA^o)	γ	$r (A^o)$	$\Delta H = \frac{P_o \gamma}{r(n^*+1)}$, (eV)	$r^u (A^o)$	$\Delta H = \frac{P_o \gamma}{r_u(n^*+1)}$, (eV)	ΔH calculation (kJ/mol)	ΔH°_{298} ref. data (kJ/mol)
	B	$2S^2\,2p^1$	16.086	1	0.91	5.892			568.7	561
3б	Al	$3S^2\,3p^1$	18.093	1	1.43 $r_k = 1.25$	3.163 3.619			<327.3>	329.3
	Ga Ga	$4S^2\,4p^1$	20.760 8.8961	1 2	1.39 1.39	3.178 2.723			<284.8>	273.0
3б	In	$5S^2\,5p^1$	21.841	1	1.66	2.6314			<241.5>	238.1
		$5S^2\,5p^1$	21.841	½			0.92	2.374		
	Tl	$6S^2\,6p^1$	22.012	½			1.05	2.015	194.5	181.0
	C	$2p^2$ $2S^2\,2p^2$	10.061 24.585	1 1	0.77 0.77	4.3554 10.643			<723.9>	716.7
	Si	$3p^2$	10.876	2	1.17	4.648			448.6	452
4б	Ge	$4p^2$ $4p^2$	12.072 12.072	2 2	1.39 1.24	3.3656 4.1428			<378.3>	376.6
	Ge	$4p^2$	12.072	1			0.65	3.9516	381.43	376.6
	Sn	$5p^2$	13.009	2	1.58	3.2934			317.9	302.1
	Pb	$6S^2\,6p^2$	32.526	1	1.50	2.0952			202.2	195.1
	Pb	$6p^2$	13.460				1.26	2.0543	198.3	195.1

Table 4. calculation of formation enthalpy of single-atom gases – ΔH_{298}

$1\ eV = 96.525\ kJ/mol$ **(Continued)**

Group	Atom	Orbitals	P_o (eVA^o)	γ	$r(A^o)$	$\Delta H = \frac{P_o\gamma}{r(n^*+1)}$, (eV)	$r_u(A^o)$	$\Delta H = \frac{P_o\gamma}{r_u(n^*+1)}$, (eV)	ΔH calculation (kJ/mol)	ΔH° 298 ref. data (kJ/mol)
	N	$2p_r^5$	21.966				1.48	4.947	477.5	472.7
5б	P	$3p^1$	7.7864	2	r_K =1.16	3.3562			323.9	316.3
	As (5) As (3)	$4S^2 4p^3$ $4p^3$	40.232 18.645	½ 1	1.40 r_K = 1.21	3.057 3.2785			295.1 316.5	301.8
	As (3 ,5)								<305.8>	
5б	Sb	$5p^3$ $5p^3$	20.509 20.509	1 1	r_K = 1.39 1.61	2.9509 2.5477			<265.2>	268.2
	Bi	$6p^3$	21.919	1	1.82	2.3160			<207.3>	209.2
	Bi	$6p^3$	21.919	1			2.13	1.9790		
	O (2)	$2p^2$	11.858	1			1.40	2.823	<253.1>	249.2
	O (4)	$2p^4$	20.338	½			1.40	2.421		
6б	S (2)	$3S^1 3p^1$	21.673	2	r_K =0.94	2.882			<280.8>	277.0
	S (4)	$3p^4$	21.375	1			1.82	2.9360		
	Se (4)	$4p^4$	24.213	½	r_K =.17	2.2032			<226.7>	223.4
	Se (2)	$4S^1 4p^1$	23.283	1			1.98	2.495		

Group	Atom	Orbitals	P^o (eVA^o)	γ	$r(A^o)$	$\Delta H = \frac{P_o\gamma}{r(n^*+1)}$, (eV)	$r^u(A^o)$	$\Delta H = \frac{P_o\gamma}{r_u(n^*+1)}$, (eV)	ΔH calculation (kJ/mol)	$\Delta H^\circ 298$ ref. data (kJ/mol)
	Te	$5S^1 5p^1$	23.882	1			2.11	2.264	218.5	215.6
	Po	$6S^1 6p^1$	23.664	½	$r_K = 1.50$	1.516			146.4	146
7б	F	$2p^1$	6.635	½			1.33	0.83145	80.3	79.5
	Cl	$3p^1$	8.5461	1			1.81	1.2804	123.6	121.3
	Br	$4p^1$	9.3068	1			1.79	1.1062	106.8	111.8
	I	$5p^1$	9.9812	½			r^{5+} = 0.94	1.062	102.5	106.8

The heats of formation of chemical elements in standard state are taken as zero. The list of such elements is known, e.g. [13] and mainly incorporates elements in solid state. Thus the single-atom gas formation enthalpy in these cases equals the sublimation enthalpy and is determined by physical and chemical criteria of the process. The sublimation comes to diffusion movements of heated particles into the surface layer with further effusion (outflow). The diffusion activation energies in external and internal regions are quite different [14].

It is usually believed that sublimation flows by the particle migration from more strongly bonded state with the biggest number of neighbors to less strongly bonded and further – to the adsorbed surface layer [15]. By analogy, the number of interacting molecules in liquid decreases, when molecules move from the lower part of the surface layer to its upper part.

The diffusion activation energy is defined by the values of electron densities of migrating particle and particles surrounding it at the distance of atom-molecular interaction radius (R), i.e. such interaction is carried out via the power field of particles evaluated by the values of their P-parameters [16]. In such a model the diffusion process has basically the similar nature in any aggregate state and its energy and directedness are defined by three main factors:

1) value of P-parameters of structures;
2) number of particles;
3) radius of atom-molecular inetarction.

In advanced researches carried out for solid solutions in the frames of generalized grid model it has been found out that “effective diffusion coefficient depends on local composition, own volumes of component atoms and potentials of paired interactions” [17]. At the same time two types of diffusion are also distinguished: “normal” and “bottom-up” [18], this apparently agrees with the notion of volume (internal) and surface diffusion.

In a liquid, the radius of the sphere of molecular interaction $R \approx 3r$, where r – radius of a molecule. Liquids are mainly formed by the elements of first and second periods of the system. For the second period we can write down: $R \approx 3r = (n+1)r$, where n – main quantum number. For both periods (first and second) we have $R = (\langle n \rangle + 1)r \approx 2{,}5r$.

Let us assume that this principle with definite approximation can be extended to various aggregate states of elements of all other periods but taking into consideration screening effects introducing the value of effective main quantum number (n^*) instead of n. These values of n^* and n^*+1 taken by Slater [19] are shown in Table 3.

Thus we assume that the radius of sphere of atom-molecular interaction during the particle diffusion in the sublimation process is defined as follows:

$$R = (n^* + 1)r,$$

where r – dimensional characteristic of atom structure. Total change of R is from 3r to 5,2r.

When forming single-atom gases, the sublimation process is accompanied by the rupture of paired bond of atoms of near surroundings. The averaged value of structural P_S-parameter of interacting atoms can be the assessment of formation enthalpy and numerically equals the value of P_0-parameter falling on the radius unit of atom-molecular interaction but taking into consideration the relative number of interacting particles by the equation:

$$P_S = \frac{P_0}{R}\gamma = \frac{P_0\gamma}{r(n^*+1)} \approx \Delta H^o_r$$

where γ - coefficient taking into consideration the relative number of interacting particles and equaled to (as the calculations demonstrated):

$$\gamma = \frac{N_0}{N}$$

Here N_0 – number of particles in the sphere volume of the radius R,

N – number of particles of realized interactions depending on the process type (internal or surface diffusion).

Inside the liquid below the top layer 2R thick, the resultant force of molecular interaction equals zero.

Applying the initial analogy to internal diffusion and sublimation we can consider that such equilibrium state corresponds to the equality N_0=N, and then γ=1.

On the top part of liquid surface layer the volume of the sphere of atom-molecular interaction and the number of particles in it is practically 2 times lower in comparison with internal layers under 2R, i.e. $\frac{N_0}{N} \approx \frac{1}{2}$ and $\gamma=\frac{1}{2}$ – for surface diffusion and sublimation.

In case of volume diffusion even a more extreme option is possible, when the number of particles of realized interactions 2 times exceeds N_0.

Thus two-valent elements magnesium and calcium form either two or even four covalent bonds (in chelate compounds): two covalent bonds and two – by donor-acceptor mechanism. For such and analogous cases with volume diffusionγ=2.

Calculations and Comparisons

Based on such initial statements and assumptions the value of P_S-parameter was calculated (and ΔH^0_r) by the equation (12) for the majority of elements in periodic system – Table 4. The values of covalent, atom and ion radii are basically taken by Belov-Bokiy and partially by Batsanov [20].

The structural coefficient γ was taken as equaled to 2 for covalent spatial-energy bonds at the distances of only atom or covalent radii (surface diffusion). Coefficient γ was taken as equaled to 1 with ion spatial-energy bonds of elements from subgroups «a» and group 8 in periodic system (internal diffusion). In all other cases this coefficient equals 1 or 1/2 (internal or volume diffusion).

In some cases the values of ΔH^0_r of the given element was found as average value by its two possible valent states (marked with <...>). The calculations are carried out practically for all elements of six periods irrespectively of their aggregate state that definitely could give higher accuracy than reference data.

Experimental and reference data [2, 3] have a relative error within 0.5 – 1.5 (%), but in the given calculations such average error was about 5 %. Probably the search for a more rational technique of registering screening effects for clarifying the effective main quantum number can eventually bring more reliable results for evaluating ΔH^0_r.

Conclusions

The dependences of dissociation energy of binary molecules and formation enthalpies of single-atom gases upon initial spatial-energy characteristics of free atoms have been determined.

The corresponding equations, the calculations on which basically agree with the experimental and reference data, have been obtained.

References

[1] Lebedev Yu.A., Miroshnichenko E.A. Thermal chemistry of steam-formation of organic substances. – M.: *"Nauka",* 1981, 216 p.

[2] Properties of inorganic compounds. Reference-book/ / Efimov A.I. et al. – L.: *"Khimiya",* 1983, 392 p.

[3] Rupture energy of chemical bonds. Ionization potentials and affinity to an electron. Reference-book/ / Kondratyev V.I. et al. – M.: *"Nauka",* 1974, 351p.

[4] Benson S. Thermochemical kinetics. – M.: *Mir,* 1971, 308 p.

[5] Berlin A.A., Wolfson S.A., Enikolopyan N.S. Kinetics of polymerization processes. – M.: *"Khimiya",* 1978, 319 p.

[6] Korablev G.A. *Spatial-Energy Principles of Complex Structures Formation,* Netherlands, Brill Academic Publishers and VSP, 2005,426p. (Monograph).

[7] Korablev G.A. *Spatial-energy criteria of phase-formation processes.* – Publishing house "Udmurt university", Izhevsk, 2008, 494 p.

[8] Fischer C.F. Average-Energy of Configuration Hartree-Fock Results for the Atoms Helium to RadonV//*Atomic Data,*-1972, -№ 4, -p. 301-399.

[9] Waber J.T., Cromer D.T. Orbital Radii of Atoms and Ions//*J. Chem. Phys* -1965, -V 42, -№12, -p. 4116-4123.

[10] Clementi E., Raimondi D.L. Atomic Screening constants from S.C.F. Functions, 1.//J.*Chem. Phys*.-1963, -v.38, -№11, -p. 2686-2689

[11] Clementi E ., Raimondi D.L. Atomik Screening Constants from S.C.F. Functions, 2.//*J. Chem. Phys.*- 1967,- V.47, № 4,- p. 1300 – 1307.

[12] Korablev G.A., Zaikov G.E. Energy of chemical bond and spatial-energy principles of hybridization of atom orbitalls.//*J. of Applied Polymer Science.* USA, 2006., V. 101, № 3, p. 283 – 293.

[13] Thermodynamic properties of individual substances. Reference-book of Academy of Science of the USSR, ed. Glushko V.P., M.: *"Science",* 1978, v.1, 496 p.

[14] Kofstad P. Deviation from stoichio, diffusion and electrical conductivity in simple metal oxides. M.: *"Mir",* 1975, 398 p.

[15] Pound G.M. – S. Phys. and Chem. Ref. Data. 1972, v.1. p. 135 – 146.

[16] Korablev G.A., Solovyev S.D. *Energy of diffusion activation in metal* systems//Bulletin of ISTU, 2007, №4, p.128-132.

[17] Zaharov M.A. Grid models of multi-component solid solutions: statistic thermodynamics and kinetics//*Abstract of doctoral thesis.* Novgorod State University, 2008, 36 p.

[18] Geguzin Ya.E. *Ascending diffusion and diffusion aftereffect//UFN.*-1986, v.149, iss.1, p. 149-159.

[19] Batsanov S.S., Zvyagina R.A. Overlap integrals and problem of effective charges. “Nauka”, Siberian branch, Novosibirsk, 1966, 386 p.

[20] Batsanov S.S. Structural chemistry. *Facts and dependencies.* – M.: MSU – 2000, 292 p.

In: Chemical and Biochemical Kinetics: New Perspectives ISBN 978-1-61209-224-9
Editor: Gennady Zaikov

Chapter 3

DEPENDENCE OF NANOSTRUCTURE FORMATION DEGREE UPON SPATIAL-ENERGY CHARACTERISTICS OF FREE ATOMS

G. A. Korablev,[1,*] V. I. Kodolov[1,†] and G. E. Zaikov[2,#]
[1]BasicResearch-EducationalCenter of Chemical Physics and Mesoscopy,
Udmurt Research Center, Ural Division, Russian Academy of Science, Izhevsk, Russia
[2]N.M. Emmanuel Institute of Biochemical Physics,
Russian Academy of Science, Moscow, Russia

ABSTRACT

It is determined that the rate of carbonization of nanostructures coincide graphically with the nomogram of the dependence of structural interactions relative difference of P-parameters.

The graphical dependence of time of nanostructure formation upon the initial energy characteristics of atoms, coinciding with the nomogram, is obtained in the model of surface diffusive process.

Keywords: nanostructures, carbonization rate, spatial-energy parameter, diffusion.

INTRODUCTION

The problem of *a priori* evaluation of energy conditions of the formation rate of stable nanostructures is one of the challenges in molecular nanotechnology. Its solution, in turn, is directly connected with the regularities of nanostructure interactions and phase-formation in general. Apparently the immediate solution to such problems can be only based on

[*] E-mail: korablev@udm.net.
[†] E-mail: kodol@istu.udm.net.
[#]E-mail: chembio@sky.chph.ras.ru.

fundamental principles defining the aggregate of physical-chemical criteria of a substance and such basis is possessed by quantum-mechanical concepts of physics and chemistry of solids.

But a lot of calculations of phase-formation based on the application of quantum-mechanical techniques and statistic-thermodynamic theories are still made only for a relatively small number of systems. Many papers dedicated to the phenomenon of structure formation, creation of an adequate model of nanoreactors, energy theories of solid solutions are being published. But still a lot issues in theoretical and prognosticating evaluation of phase-formation of nanosize systems are not solved.

The method utilizing complex initial characteristics of an atom as the stabilization criterion of nanostructures is developed in this paper.

SPATIAL-ENERGY PARAMETER (P-PARAMETER)

Lagrangian equation for the relative motion of isolated system of two interacting material points with masses $m_1$and$m_2$in coordinate x with acceleration α can be written down as follows:

$$\frac{1}{1/(m_1 a\Delta x)+1/(m_2 a\Delta x)} \approx -\Delta U \quad \text{or} \quad \frac{1}{\Delta U} \approx \frac{1}{\Delta U1} + \frac{1}{\Delta U2} \tag{1}$$

where ΔU_1 and ΔU_2 – potential energies of material points on the elementary segment of interactions, ΔU – resultant (mutual) potential energy of these interactions.

The atoms system is formed from differently charged masses of the nucleus and electrons. In this system the energy characteristics of subsystems are: orbital energy of electrons (W_i) and effective energy of the nucleus taking into account screening effects (by Clementi).

Therefore assuming that the resultant interaction energy in the system orbital-nucleus (responsible for interatom interactions) can be calculated by the principle of adding reverse values of some initial energy components, the introduction of P-parameter [5] as an averaged characteristic of valence orbitals was substantiated based on the following equations:

$$\frac{1}{q^2/r_i} + \frac{1}{W_i n_i} = \frac{1}{P_E} \tag{2}$$

$$P_E = \frac{P_0}{r_i} \tag{3}$$

$$\frac{1}{P0} = \frac{1}{q^2} + \frac{1}{(Wrn)_i} \tag{4}$$

$$q = \frac{Z^*}{n^*} \tag{5}$$

Here: W_i – bond energy of electrons [6]; r_i– orbital radius of i orbital [7]; Z^* and n^* – effective charge of the nucleus and effective main quantum number [8].

P_0 will be designated as spatial-energy parameter, and P_E – effective P-parameter. Effective P_E-parameter has a physical sense of some averaged energy of valence electrons in the atom and is measured in energy units, e.g. electron-volts (eV).

According to the calculations P_E-parameters numerically equal (in the range of 2%) the total energy of valence electrons (U) by the atom statistic model. Using the known ratio between the electron density (β) and interatom potential the direct dependence of P_E-parameter upon the electron density at the distance r_i from the nucleus can be obtained by the atom statistic model:

$$\beta_i^{2/3} = \frac{AP_0}{r_i} = AP_E$$

where A – constant.

The reliability of this equation was confirmed by the calculations of electron density using wave functions by Clementi and comparing it with the value of electron density calculated via the value of P_E-parameter.

The introduction of P-parameter as a criterion of structural interactions is based on the assumption that the resultant energy in the system orbital-nucleus immediately responsible for interatom interactions can be calculated by the principle of adding reverse values of some initial atom characteristics in the initial state. In this model P_0-parameter is the tabulated constant spatial-energy characteristic of each atom orbital. P_0-parameter is calculated for the majority of elements in the periodic system [5], some are given in Table 1.

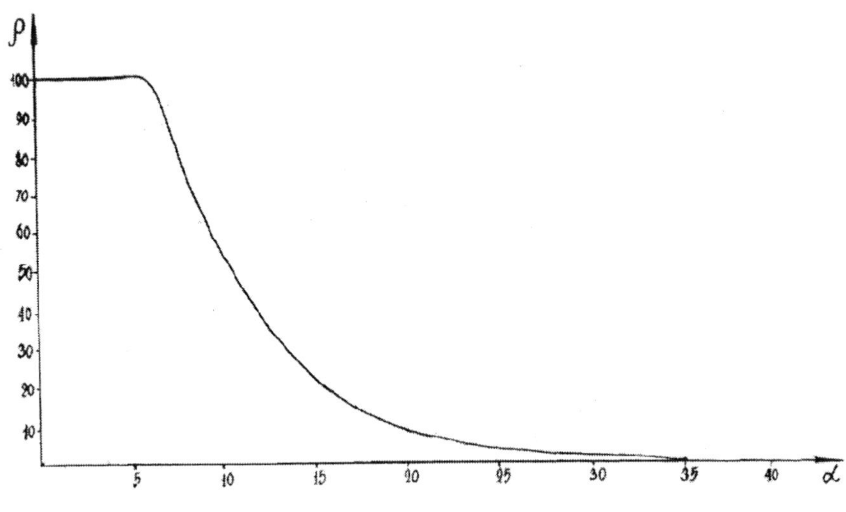

Figure 1a. Nomogram №1.

Spatial-energy principles of isomorphic substitution were found:

1. Complete isomorphic substitution takes place at approximate equality of P-parameters of valence orbital of interchanging atoms: $P_E^{'} \approx P_E^{''}$.

The degree of isomorphic interactions is determined through the relative difference of P_E-parameters of atoms-components by the following formula:

$$\alpha = \frac{P_E^{'} - P_E^{''}}{P_E^{'} + P_E^{''}} 200\% \qquad (5)$$

2. P-parameters of valence orbital of the atom of its smallest value define the orbital that is mainly responsible for isomorphism and structural interactions.

But isomorphism is the particular case of phase-formation. Therefore taking its ratios as the basis, it can be assumed that during the formation of crystalline structure the degree of structural interactions can be characterized via the relative difference of P-parameters (coefficient α).

Application of the ratios (2-5) to different types of crystalline structures using tabulated values of P_0-parameters demonstrated the rational possibility of such an approach for the evaluation of energy conditions of nanostructure formation.

Multiple calculations for simpler and more complicated systems of different structural types allowed preparing a unified nomogram of the dependence of the degrees of structural interactions (ρ) upon the coefficient α (Figure 1a):$\rho_1 = f(\alpha_1)$

Graphical dependencies of the degree of activities or formation of substrates of different bio-physical-chemical process upon kinetic and energy characteristics are similar. For example, the dynamics of electron-conformation interactions [9], frequency dependence of electric resistance of tissues, distribution of energies of channel conductivity fluctuations as frequency functions [9], etc.

RESEARCH TECHNIQUE

As an example we take the process of carbonization and nanostructure formation during the interactions in gels of polyvinyl alcohol and metal phase in the form of oxides or copper chlorides [10].On the first stage small clusters of inorganic phase surrounded by organic products are formed. In this period the main character of atom-molecular interactions should be evaluated via the relative difference of P-parameters calculated via the radii of copper ions and covalence radii of carbon atoms given in Tables 1, 2:

From Table 2 it is seen that in this case the coefficient α = 3.50 corresponding to the complete structural interaction: ρ = 100%. The process flows only in the gel volume, not on the surface of the film not formed yet.

In the following basic period of carbonization the metal phase is formed on the surface of polymeric structures being formed by the following reaction [10]:

$2CuCl + [-CH{=}CH-]_n \rightarrow 2Cu + 2HCl + [C_2]_n$

Table 1. P-parameters of atoms calculated via the bond energy of electrons

Atom	Valence electrons	W (eV)	r_i (Å)	q^2_0 (eVÅ)	P_0 (eVÅ)	R (Å)	P_0/R (eV)	$P_0/R(n^*+1)$	$r_и$ (Å)	P_0/r_i (eV)	$P_0/R_i(n^*+1)$
C	$2p^1$	11.792	0.596	35.395	5.8680	0.77	7.6208	2.5403	2.60	2.2569	0.75231
	$2p^2$	11.792	0.596	35.395	10.061	0.77	13.066	4.3554	2.60	3.8696	1.2899
	$2S^1$	19.20119.201	0.620	37.240	9.0209	0.77	11.715	3.9052	2.60	3.4696	1.1565
	$2S^2$		0.620	37.240	14.524	0.77	18.862	6.2874	2.60	5.5862	1.8621
	$2p^2+2S^2$				24.585	0.77	31.929	10.643	2.60	9.4558	3.1519
Cu	$4S^1$	7.7485	1.191	30.117	7.0964	1.28	5.5438	1.1795	0.98	7.2408	1.5406
	$4S^2$				11.444		8.9406	1.9023	0.80	14.305	3.0436
	$3d^1$	20.142	0.312	278.78	6.1457		4.8013	1.0216			
	$3d^2$				12.026						
	$4S^1+3d^1$				13.242	1.28	10.345	2.2011	0.80	16.553	3.5218
	$4S^1+3d^2$				19.122		14.939	3.1786			
	$4S^2+3d^1$				17.590		13.742	2.9238			

Table 2. Structural interactions during the formation of nanofilm in the system C→CU

Carbon atom				Copper atom			Interaction features				
Orbital	$\frac{P_o}{R(n^*+1)}$ (eV)	$\frac{P_o}{r_i(n^*+1)}$ (eV)	K	Orbital	$\frac{P_o}{R(n^*+1)}$ (eV)	$\frac{P_o}{r_i(n^*+1)}$ (eV)	α (%)	ρ (%)	t (hour)	ω (%)	Type of interactions
$2P^22S^2$		3.1519	1	$4S^2$		3.0436	3.50	100	0	0	volumetric
$2P^2$	4.3554		1.6	$4S^1 3d^1$	2.2011		21.17	5-8	0	0	semi-surface
$2P^2$	4.3554		1.7	$4S^1 3d^1$	2.2011		15.15	19-21	0.49	21.5	surface
$2P^2$	4.3554		1.8	$4S^1 3d^1$	2.2011		9.46	56-58	1.05	63.9	surface
$2P^2$	4.3554		1.9	$4S^1 3d^1$	2.2011		4.06	~100-98	1.6	95	surface
$2P^2$	4.3554		2.0	$4S^1 3d^1$	2.2011		1.07	100	2.0	98.3	surface

Table 3.Effective main quantum number

n	1	2	3	4	5	6
n^*	1	2	3	3.7	4	4.2
n^*+1	2	3	4	4.7	5	5.2

From this moment the process of the formation of binary matrix of the nanosystem starts. Let us consider the formation process of film carbon matrix in copper based on the surface diffusion model. At this point of the formation of metal phase P-parameters calculated via atom radii are valence active (Tables 1,2).

CALCULATIONS AND COMPARISONS

The diffusion activation energy is determined by the values of electron densities of the migrating particle and the particles surrounding it at the distance of atom-molecular radius (R), hence there is an interaction via the energy field of particles evaluated by the values of their P-parameters [11]. In such a model the diffusion process takes place under any aggregate states and its energetics and directedness are determined by three main factors:

1) value of structure P-parameters;
2) number of particles;
3) radius of atom-molecular interaction.

It is common that diffusion (e.g. during sublimation) goes by the particle migration from stronger bound state with greater number of neighbors to a weaker one, and further to the adsorbed surface layer [12]. In a similar way during the transit of molecules from the lower part of the surface layer to its upper one in a liquid the number of interacting molecules decreases. In modern investigations carried out for solid solutions in the frameworks of unified grid model it was found out that "effective coefficient of diffusion depends on the local composition, own atom volumes of the component and potentials of paired interactions" [13]. At the same time two types of diffusion are distinguished: "normal" and "rising" [14], apparently this conforms to the concepts of volume (internal) and surface diffusions. Activation energies of diffusion in external and internal areas are rather different [15].

In a liquid the radius of the sphere of molecular interaction $R \approx 3r$, where r – molecule radius. Liquids are mainly formed by the elements of first and second periods of the system. For the second period: $R \approx 3r = (n+1)r$, where n – main quantum number. For both periods (first and second) we have: $R = (\langle n \rangle + 1)r \approx 2{,}5r$.

Let us assume that this principle with a definite approximation can be used for different elements of the other periods but taking screening effects into account, introducing the value of effective main quantum number(n^*) instead of n. These values n^* and n^*+1 taken by Slater [16] are given in Table 3.

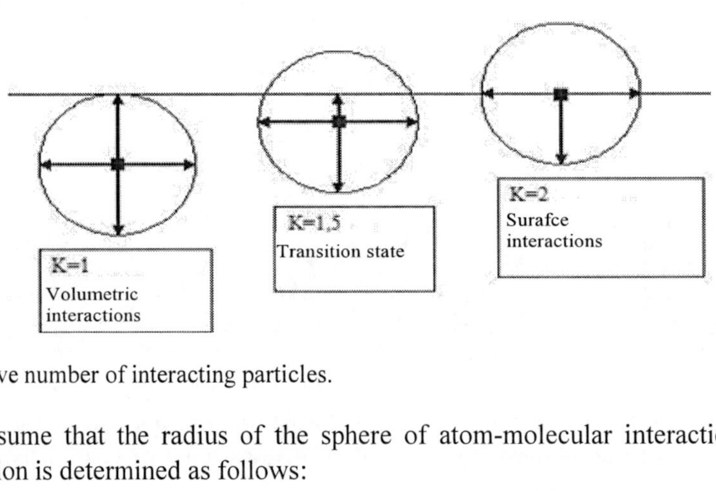

Figure 2. Relative number of interacting particles.

So we assume that the radius of the sphere of atom-molecular interaction during the particle diffusion is determined as follows:

$R=(n^*+1)r$,

where r – dimensional characteristic of atom structure. The complete change of R is from 3r to 5.2r (from second to sixth period).

The averaged value of structural P_S-parameter for the radius unit of atom-molecular interaction is determined by the following equation:

$$P_S = \frac{P_0}{KR} = \frac{P_0}{r(n^*+1)K} \qquad (7)$$

where K – coefficient considering the relative number of interacting particles and equaled to (based on calculations):

$$K = N_0 / N$$

where N_0 – number of particles in the volume of the sphere of radius R, N – number of particles or realized interactions depending upon the process type (internal or surface diffusions).

Inside the liquid below the top level 2R thick the resultant force of molecular interaction equals zero (Figure 2).

Applying the initial analogy to internal diffusion we can agree that such equilibrium state corresponds to the equality N_0=N, then K=1.

On the upper part of the liquid surface layer the volume of the sphere of atom-molecular interaction and number of particles in it are practically 2 times lower in comparison with internal layers below 2R, i.e. $\frac{N_0}{N} \approx 2$ and K=2 – for surface diffusion (Figure 2).

Actually the surface diffusion goes on at the change of coefficient "K" within the range from 1.5 to 2.0, this is taken into account in the calculations. Based on such initial conditions the value of P_S-parameter and coefficient α are calculated by the equations (5, 7) for carbon and copper atom (Table 2). The values of covalence, atom and ion radii are taken mainly from Belov-Bokiy and partly from Batsanov [17].

Using the nomogram 1 the values of the degree of structural interactions with the coefficient α are calculated, i.e. $\rho_2 = f(\alpha_2)$ – curve 3. In the figure 1 there are values of α in reverse succession compared with the Figure 1a. Here we can also see the graphical dependence of the degree of nanofilm formation (ω) upon the process time according to the data in [4] – curve 1.

The analysis of all graphical dependencies obtained shows nearly complete graphical matching of all three graphs: ω = f(t), $\rho_1 = f(\alpha_1)$, $\rho_2 = f(\alpha_2)$ with minor deviations in the beginning and end of the process. Thus the rate of carbonization can be characterized directly via the calculated values of the coefficient α.

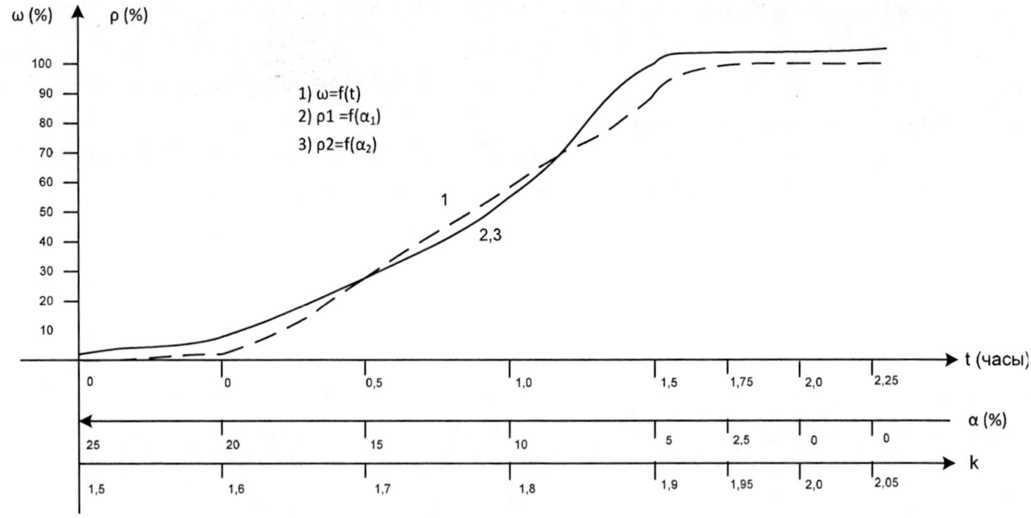

Figure 3. Dependence of carbonization rate upon the coefficient ά.

It is known [5] that nomogram 1 (Figure 1a) can be represented in the form of linear logarithmic dependence LgC= $\gamma f(\alpha)$, where γ – constant for a wide class of structures, C – composition of interacting atoms-components in mols. Therefore there is a possibility to establish the direct mathematical dependence of these parameters.

Conclusions

The rate of nanofilm formation in the system C→Cu matches the nomogram of the dependence of the degree of structural interactions upon the relative value of P-parameters of interacting atoms-components.

The possibility of evaluating the formation time of such nanofilms using primary spatial-energy parameters of free atoms is demonstrated.

References

[1] Zakharov A.Yu., Lebedev V.V. To the theory of reconstruction processes in multi-component condensed systems. // *Electron composition and properties of refractory compounds, alloys and metals.* – Proceedings of IPM NAS of Ukraine, Kiev, 2004, p. 13-21.

[2] Gusev A.I. *Nanomaterials, Nanostructions,* Nanotechnologies. M.: fizmatgiz, 2005, 412 pg.

[3] KrutikovV.A., DidikA.A., YakovlevG.I. etal. Composite material with nanoreinforcement. // *Alternative energetics and ecology,* 2005. № 4 (24). P. 36-41.

[4] Kodolov V.I., Khokhriakov N.V., Trineeva V.V., Blagodatskikh I.I. Activity of nanostructures and its display in nanoreactors of polymeric matrixes and active media. // *Chemical physics and mesoscopy*, 2008. V. 10. №4. P. 448-460.

[5] Korablev G.A. *Spatial-Energy Principles of Complex Structures Formation,* Leiden, the Netherlands, Brill Academic Publishers and VSP, 2005, 426 pages (Monograph).

[6] Fischer C.F. Average-Energy of Configuration Hartree-Fock Results for the Atoms Helium to Radon.//*Atomic Data*,-1972, -№ 4, -p. 301-399.

[7] Waber J.T., Cromer D.T. Orbital Radii of Atoms and Ions.//*J. Chem. Phys* -1965, -V 42, -№12, -p. 4116-4123.

[8] Clementi E., Raimondi D.L. Atomic Screening constants from S.C.F. Functions, 1.//*J.Chem. Phys.-1963*, -v.38, -№11, -p. 2686-2689.

[9] Rubin A.B. Biophysics. Book 1. Theoretical biophysics. Book 2. Biophysics of cellular processes, M.: *Vysshaya shkola*, 1987, 319p, 303p.

[10] Kodolov V.I., Didik A.A., Volkov A.Yu., Volkova E.G. Low-temperature synthesis of copper nanoparticles in carbon shell. // Izvestiya vuzov. *Chemistry and chemical engineering, 2004*. V.47. №1. P. 27-30.

[11] Korablev G.A., Solovyev S.D. Activation energy of diffusion in metal systems. // *Bulleting of ISTU,* 2007,№4, p.128-132.

[12] Lebedev Yu.A., Miroshnichenko E.A. Thermal chemistry of steam formation of organic particles. – M.: *Nauka,* 1981, 216 p.

[13] Zakharov M.A. Grid model of multi-component solid solutions: statistic thermodynamics and kinetics. // *Abstract of doctoral dissertation.* Novgorod State University, 2008, 36p.

[14] Geguzin Ya.E. Rising diffusion and diffusion effect. // *UFN,* 1986,v.149, iss.1, p.149-159.

[15] Kofstad P. Deviation from stoichiometry, diffusion and electrical conductivity in simple metal oxides. M.: *Mir,* 1975, 398 p.

[16] Batsanov S.S., Zvyagina R.A. Overlapping integrals and effective charge problem. *Nauka,* Siberian branch, Novosibirsk, 1966, 386 p.

[17] Batsanov S.S. Structural chemistry. *Facts and dependencies.* Moscow: MSU, 2000, 292 p.

In: Chemical and Biochemical Kinetics: New Perspectives
Editor: Gennady Zaikov
ISBN 978-1-61209-224-9

Chapter 4

QUANTUM-CHEMICAL RESEARCH OF THE MECHANISM OF SYNTHESIS OF 5-ACETYLOXYMETHYL-2-CHLORO-5-ETHYL-1,2,3-DIOXAPHOSPHORYNANE

***V. A. Babkin,*[*1] *V. U. Dmitriev,*[1] *G. A. Savin,*[2,†]**
***E. S. Titova,*[3] *and G. E. Zaikov*[4,‡]**

[1]Volgograd State Architect-Build University Sebrykov Departament, Michailovka, Volgograd region, Russia
[2]Volgograd State Pedagogical University, Volgograd; Russia
[3]Volgograd State Technical University, Volgograd; Russia
[4]Institute of Biochemical Physics Russian Academy of Sciences; Moscow, Russia

ABSTRACT

Quantum-chemical research of the mechanism of synthesis of 5-acetyloxymethyl-2-chloro-5-ethyl-1,2,3-dioxaphosphorynane for the first time is executed by classical method MNDO. Synthesis of this compound is the result of interaction of acetyl chloride and 4-ethyl-2,6,7-trioxa-1-phosphabicyclo[2.2.2]octane. Interaction proceeds on the mechanism of bimolecular nucleophylic substitution S_N2. Reaction is endothermic and has barrier character. The size of a power barrier makes 178 kJ/mol.

Keywords: quantum-chemical research, the mechanism of synthesis, 5-acetyloxymethyl-2-chloro-5-ethyl-1,2,3-dioxaphosphorynane, 4-ethyl-2,6,7-trioxa-1-phosphabicyclo[2.2.2] octane, acetyl chloride, method MNDO.

[*] E-mail: sfi@reg.avtlg.ru.
[†] E-mail: gasavin@mail.ru.
[‡] E-mail: chembio@sky.chph.ras.ru.

AIMS AND BACKGROUNDS

5-Acetyloxymethyl-2-chloro-5-ethyl-1,2,3-dioxaphosphorynane is synthesized by direct acylation of 4-ethyl-2,6,7-trioxa-1-phosphabicyclo[2.2.2]octane with acetyl chloride for the next scheme:

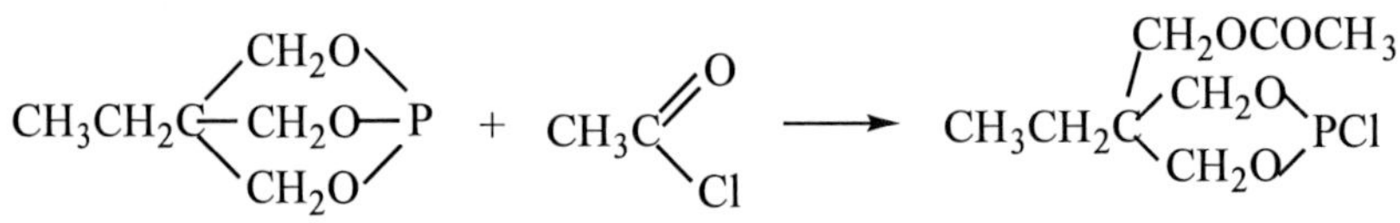

This synthesis can be rather perspective for reception of various new chemical compounds and, in particular, medical products with useful and probably unique properties. The mechanism of synthesis of 5-acetyloxymethyl-2-chloro-5-ethyl-1,2,3-dioxaphosphorynane now is not studied. Research of the mechanism of synthesis of this compound was the purpose of the present work. Research was spent by one of classical quantum-chemical methods.

METHODICAL PART

The Quantum-chemical semi empirical method MNDO with optimization of geometry on all parameters by the standard gradient method which has been built in PC GAMESS has been chosen for research of the mechanism of synthesis of 5-acetyloxymethyl-2-chloro-5-ethyl-1,2,3-dioxaphosphorynane [2]. This method well enough reproduces power characteristics and stability of chemical compounds, including the substances containing multiple bonds [1]. Calculations were carried out in approach to the isolated molecule to a gas phase. The Program MacMolPlt was used for visual representation initial, intermediate and final models [3].

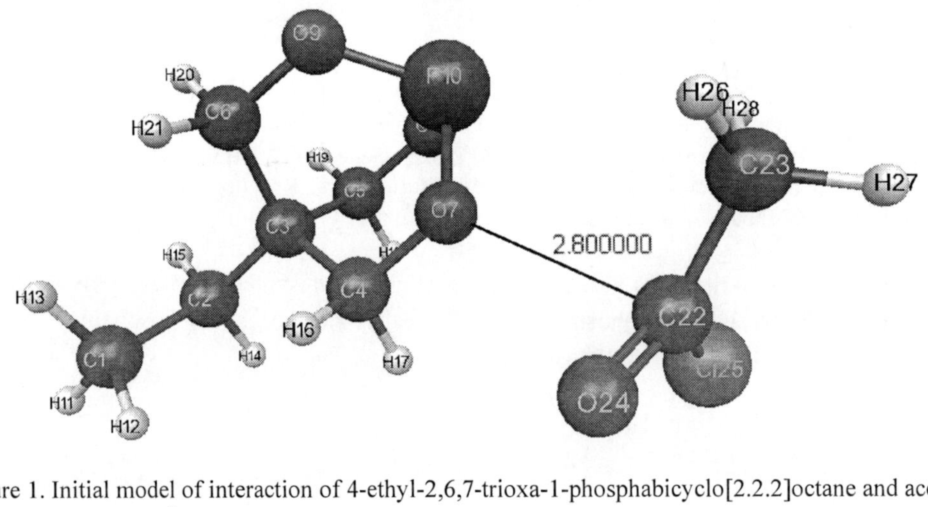

Figure 1. Initial model of interaction of 4-ethyl-2,6,7-trioxa-1-phosphabicyclo[2.2.2]octane and acetyl chloride. $R_{O7C22} = 2.8Å$.

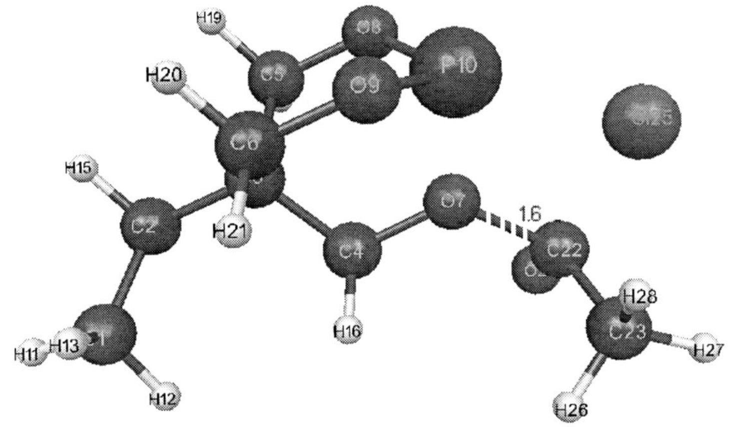

Figure 2. Model of a stage of break of bonds (a transitive condition). R_{O7C22} = 1.6Å.

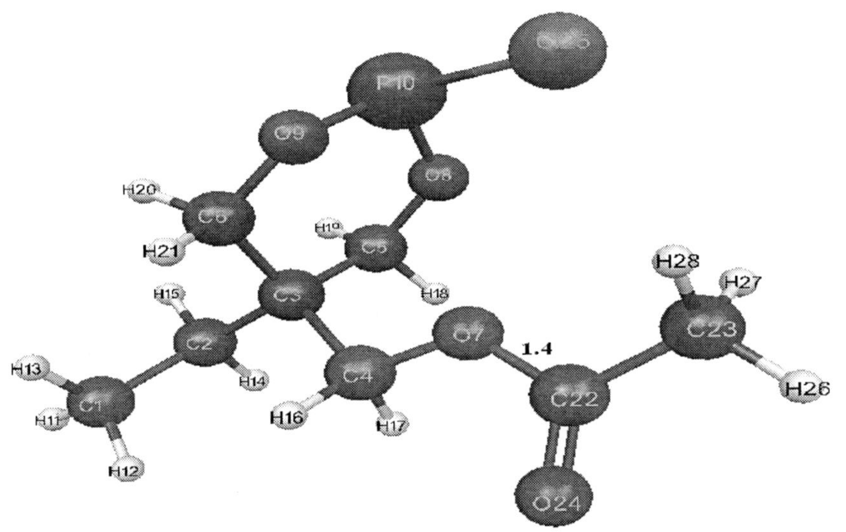

Figure 3. Final model - formation of 5-acetyloxymethyl-2-chloro-5-ethyl-1,2,3-dioxaphosphorynane. R_{O7C22} = 1.4Å.

The Mechanism of synthesis of studied compound was investigated by method MNDO. Initial components of synthesis – 2,6,7-trioxa-4-ethyl-1-phosphabicyclo[2.2.2]octane and acetyl chloride – settled down on distance 2.8-3.0Å from each other. On such distance any interactions between them practically are absent. As coordinate of reaction distance R_{O7C22} (figure 1) has been chosen. This distance is most energetically a favorable direction of interaction of initial components by the maximal values of charges on atoms of molecular system P_{10},O_7,Cl_{25} and C_{22} directly participating in reaction. Optimization on all parameters of initial components at R_{O7C22} = 2.8Å was carried out. Values of lengths of bonds and valent corners, values E_0 (the general energy of system) and q_H-a charge on atoms along coordinate of reaction R_{O7C22} were fixed and brought in Tables 1-3 after optimization. The coordinate of reaction changed with 2.8Å up to 1.2Å at each step of optimization. The step on coordinate of reaction R_{O7C22} has made 0.2Å.

Table 1. Change of energy along coordinate of reaction R_{O7C22}

R_{O7C22}, Å	E_0, kJ/mol
2,8	-291546
2,6	-291530
2,4	-291507
2,2	-291475
2,0	-291433
1,8	-291394
1,6	-291365
1,4	-291544
1,3	-291533
1,2	-291444

Table 2.Change of lengths of bonds along coordinate of reaction R_{O7C22}

R_{C22-O7}, Å	R_{C1-C2}	R_{C2-C3}	R_{C3-C4}	R_{C3-C5}	R_{C3-C6}	R_{C4-O7}	R_{C5-O8}	R_{C6-O9}	R_{P10-O8}	R_{P10-O7}
2.8	1.53	1.55	1.58	1.58	1.58	1.40	1.39	1.39	1.60	1.61
2.6	1.53	1.55	1.58	1.58	1.58	1.40	1.39	1.39	1.60	1.62
2.4	1.53	1.55	1.58	1.58	1.58	1.40	1.39	1.39	1.60	1.62
2.2	1.53	1.55	1.58	1.58	1.58	1.40	1.39	1.39	1.60	1.63
2.0	1.53	1.55	1.58	1.58	1.58	1.41	1.39	1.39	1.60	1.64
1.8	1.53	1.55	1.57	1.58	1.58	1.42	1.39	1.39	1.60	1.65
1.6	1.53	1.55	1.57	1.58	1.58	1.43	1.39	1.39	1.59	1.69
1.4	1.53	1.57	1.57	1.58	1.57	1.40	1.38	1.38	1.58	3.62
1.3	1.53	1.57	1.57	1.58	1.57	1.41	1.38	1.38	1.59	3.66
1.2	1.53	1.57	1.57	1.58	1.57	1.42	1.38	1.38	1.59	3.71

R_{C22-O7}, Å	$R_{P10-Cl25}$	R_{H11-C1}	R_{H12-C1}	R_{H13-C1}	R_{H14-C2}	R_{H15-C2}	R_{H16-C4}	R_{H17-C4}
2.8	4.43	1.11	1.10	1.10	1.11	1.11	1.11	1.11
2.6	3.59	1.11	1.10	1.10	1.11	1.11	1.11	1.11
2.4	3.39	1.11	1.10	1.10	1.11	1.11	1.11	1.11
2.2	3.19	1.11	1.10	1.10	1.11	1.11	1.11	1.11
2.0	3.03	1.11	1.10	1.10	1.11	1.11	1.11	1.11
1.8	2.86	1.11	1.10	1.10	1.11	1.11	1.11	1.11
1.6	2.63	1.11	1.10	1.10	1.11	1.11	1.11	1.11
1.4	2.02	1.11	1.10	1.10	1.11	1.11	1.12	1.12
1.3	2.02	1.11	1.10	1.10	1.11	1.11	1.12	1.12
1.2	2.02	1.11	1.10	1.10	1.11	1.11	1.11	1.12

Table 2. Continued

R_{C22-O7},Å	R_{H18-C5}	R_{H19-C5}	R_{H20-C6}	R_{H21-C6}	$R_{C23-C22}$	$R_{O24-C22}$	$R_{Cl25-C22}$	$R_{H26-C23}$	$R_{H27-C23}$	$R_{H28-C23}$
2.8	1.11	1.11	1.11	1.11	1.51	1.20	1.80	1.10	1.10	1.10
2.6	1.11	1.11	1.11	1.11	1.51	1.20	1.80	1.10	1.11	1.10
2.4	1.11	1.11	1.11	1.11	1.51	1.20	1.81	1.10	1.11	1.10
2.2	1.11	1.11	1.11	1.11	1.52	1.20	1.82	1.10	1.11	1.10
2.0	1.11	1.11	1.11	1.11	1.52	1.21	1.84	1.10	1.11	1.10
1.8	1.11	1.11	1.11	1.11	1.53	1.21	1.87	1.10	1.11	1.10
1.6	1.11	1.11	1.11	1.11	1.54	1.22	1.95	1.10	1.10	1.10
1.4	1.12	1.12	1.12	1.12	1.51	1.22	4.42	1.10	1.10	1.10
1.3	1.12	1.12	1.12	1.12	1.53	1.23	4.43	1.10	1.10	1.10
1.2	1.12	1.12	1.12	1.12	1.54	1.24	4.45	1.10	1.10	1.10

Table 3. Change of charges of system along coordinate of reaction R_{O7C22}

R_{C22-O7}, Å	2,8	2,6	2,4	2,2	2,0	1,8	1,6	1,4	1,3	1,2
C1	0.02	0.02	0.02	0.02	0.02	0.02	0.02	0.02	0.02	0.02
C2	0.02	0.02	0.02	0.02	0.02	0.02	0.01	0.00	0.00	0.00
C3	-0.23	-0.23	-0.23	-0.23	-0.23	-0.23	-0.22	-0.17	-0.17	-0.17
C4	0.20	0.20	0.20	0.19	0.19	0.19	0.19	0.21	0.22	0.23
C5	0.20	0.20	0.20	0.20	0.20	0.20	0.20	0.23	0.23	0.22
C6	0.20	0.20	0.20	0.20	0.20	0.20	0.20	0.23	0.23	0.23
O7	-0.52	-0.53	-0.54	-0.53	-0.53	-0.51	-0.45	-0.35	-0.32	-0.28
O8	-0.50	-0.50	-0.50	-0.50	-0.50	-0.51	-0.51	-0.51	-0.51	-0.51
O9	-0.51	-0.51	-0.52	-0.52	-0.51	-0.52	-0.53	-0.52	-0.52	-0.51
P10	0.98	1.00	1.00	1.02	1.03	1.07	1.14	0.98	0.98	0.97
H11	0.00	0.00	0.00	0.00	0.00	0.01	0.01	0.00	0.00	0.01
H12	0.00	0.00	0.00	0.00	0.00	0.00	0.00	0.00	0.00	0.00
H13	0.00	0.00	0.00	0.00	0.00	0.00	0.00	0.00	0.00	0.00
H14	0.01	0.01	0.01	0.01	0.01	0.01	0.02	0.01	0.01	0.01
H15	0.01	0.01	0.01	0.01	0.01	0.01	0.02	0.01	0.01	0.01
H16	0.03	0.02	0.02	0.01	0.02	0.03	0.03	0.01	0.02	0.03
H17	0.03	0.03	0.03	0.03	0.04	0.05	0.04	0.01	0.02	0.03
H18	0.02	0.02	0.02	0.02	0.03	0.03	0.03	0.03	0.03	0.04
H19	0.02	0.02	0.02	0.02	0.02	0.02	0.03	-0.01	0.00	0.00
H20	0.02	0.02	0.02	0.02	0.02	0.02	0.03	0.00	0.00	0.00
H21	0.02	0.02	0.02	0.02	0.02	0.02	0.02	0.03	0.04	0.04
C22	0.34	0.35	0.36	0.38	0.4	0.43	0.45	0.36	0.35	0.35
C23	0.02	0.03	0.03	0.03	0.03	0.03	0.03	0.05	0.04	0.04
O24	-0.23	-0.23	-0.24	-0.25	-0.28	-0.33	-0.39	-0.35	-0.39	-0.44
Cl25	-0.24	-0.24	-0.25	-0.27	-0.29	-0.34	-0.44	-0.38	-0.37	-0.36
H26	0.03	0.03	0.03	0.03	0.02	0.02	0.01	0.02	0.02	0.03
H27	0.03	0.03	0.03	0.03	0.03	0.03	0.04	0.04	0.03	0.02
H28	0.04	0.04	0.04	0.03	0.03	0.03	0.02	0.04	0.03	0.03

Table 4.Change of valencecorners along coordinate of reaction R_{O7C22}

Coordinate of reactionR_{O7C22},Å	2.8	2.6	2.4	2.2	2.0	1.8	1.6	1.4	1.3	1.2
C(3)C(2)C(1)	119	119	119	119	119	119	119	119	119	119
C(4)C(3)C(2)	113	113	113	112	112	112	112	108	107	107
C(5)C(3)C(2)	109	109	109	109	109	109	108	106	106	106
C(6)C(3)C(2)	113	113	113	113	113	112	112	109	109	109
O(7)C(4)C(3)	109	109	110	110	110	110	109	111	111	111
O(8)C(5)C(3)	109	109	109	109	109	109	109	115	115	114
O(9)C(6)C(3)	109	109	109	109	109	108	108	115	114	114
P(10)O(8)C(5)	120	120	120	121	121	121	122	127	126	126
H(11)C(1)C(2)	109	109	109	109	109	109	109	109	109	109
H(12)C(1)H(11)	107	107	107	107	107	107	107	106	106	107
H(13)C(1)C(11)	107	107	107	107	107	107	107	107	107	107
H(14)C(2)C(1)	108	108	108	108	108	108	108	108	108	108
H(15)C(2)H(14)	105	105	105	105	105	105	105	105	105	105
H(16)C(4)O(7)	108	108	108	108	108	108	108	110	109	108
H(17)C(4)H(16)	106	106	106	106	107	107	107	106	107	107
H(18)C(5)O(8)	109	109	109	109	109	109	109	107	107	107
H(19)C(5)H(18)	106	106	106	106	106	106	106	106	106	106
H(20)C(6)O(9)	109	109	109	109	109	109	109	108	108	108
H(21)C(6)H(20)	106	106	106	106	106	106	106	106	106	106
C(22)O(7)C(4)	117	122	123	123	123	123	122	123	125	129
C(23)C(22)O(7)	96	95	96	97	99	103	108	113	114	116
O(24)C(22)O(7)	128	128	128	128	128	126	125	127	124	120
Cl(25)C(22)O(7)	112	112	112	111	110	108	104	59	59	58
H(26)C(23)C(22)	111	111	111	111	111	111	110	110	110	110
H(27)C(23)H(26)	108	108	108	108	108	108	108	108	108	108
H(27)C(23)H(26)	108	108	108	108	108	108	108	108	108	108

RESULTS AND DISCUSSION

Optimized geometrical parameters of initial compounds, an intermediate condition of system (in the moment of separation of atom Cl_{25}) and the final generated condition of a molecule of 5-acetyloxymethyl-2-chloro-5-ethyl-1,2,3-dioxaphosphorynane are presented to the moment in Figures 1-3. Change of lengths of bonds, valence angles, the general energy of all molecular system and charges on atoms along coordinate of reaction R_{O7C22} with step 0.2Å are shown in Tables 1-3 and on diagrams Figures 4-8. The specified parameters testify that essential changes with system of initial components except for mutual orientation from each other at steps 1-6 (R_{O7C22} changes from 2.8Å up to 1.8Å) not occur. In this connection we shall define this stage as a stage of coordination. At II stage of interaction (7-8 steps, RO7C22 change from 1.8Å up to 1.4Å) there is a simultaneous and almost full break of two bonds C_{22}-Cl_{25} and P_{10}-O_7. Lengths of bonds accordingly change from 1.95Å up to 4.42Å and from 1.6Å up to 3,62Å accordingly (table 2). We shall define this stage, as a stage of break of bonds. Thus atom Cl_{25} of acetyl chloride attacks atom P_{10} of 4-ethyl-2,6,7-trioxa-1-phosphabicyclo[2.2.2]octane, forming covalent bond P_{10}-Cl_{25} (2.02Å). Atom O_7 and C_{22} form covalent bond O_7C_{22} (R_{O7C22} = 1.4Å). Formation of a required product comes to an end at last

III stage. We shall name it's the final stage of synthesis of 5-acetyloxymethyl-2-chloro-5-ethyl-1,2,3-dioxaphosphorynane.

It was important to Us to track change of charges on atoms directly participating in reaction P_{10}, O_7, C_{22} and Cl_{25}, and also change E_0 (the general energy of molecular system) along coordinate of reaction R_{O7C22} (Figures 4-8 and Tables 1).

Reaction has barrier character (Figure 4). The size of a barrier makes 178 kJ/mol. Thus positive charges on atoms P_{10} and C_{22} (Figures 5 and 7) change direct proportion to change E_0. Charges reach the maximal values during the moment of break of bonds C_{22}-Cl_{25} and P_{10}-O_7. Negative charges on atoms O_7 and Cl_{25} change inversely to change E_0. Negative charges also reach the maximal values (on the module) during the moment of break of the same bonds. We analyze all stated above: behavior of atoms directly participating in studied synthesis P_{10}, O_7, C_{22} and Cl_{25}, change of charges on these atoms and power of reaction - it is easy to notice, that the mechanism of synthesis of 5-acetyloxymethyl-2-chlorine-5-ethyl-1,2,3-dioxaphosphorynan represents the coordinated process with simultaneous break of bonds P_{10}-O_7 and C_{22}-Cl_{25} and formation of new bonds P_{10}-Cl_{25} and C_{22}-O_7 and carrying all features of S_N2 nucleophylic substitution.

5-Acetyloxymethyl-2-chloro-5-ethyl-1,2,3-dioxaphosphorynane is synthesized by Us for the first time and the mechanism of its synthesis is studied by quantum-chemical semi empirical method MNDO. By us it is shown, that synthesis of this compound is the result of the coordinated interactions of acetyl chloride and 4-ethyl-2,6,7-trioxa-1-phosphabicyclo [2.2.2]octane under scheme S_N2 nucleophylic substitution. It is established, that this reaction is endothermic. It will qualitatively be coordinated with experiment and has barrier character. Size of a power barrier of studied reaction makes 178 kJ/mol.

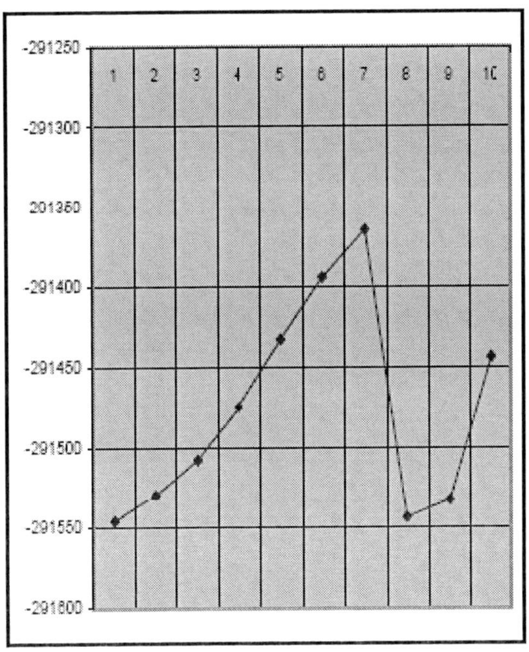

№ Step	R_{O7C22}, Å	E_0, kJ/mol
1	2,8	-291546,1384
2	2,6	-291530,4007
3	2,4	-291506,79415
4	2,2	-291475,31875
5	2,0	-291433,35155
6	1,8	-291394,0073
7	1,6	-291365,15485
8	1,4	-291543,51545
9	1,3	-291533,02365
10	1,2	-291443,84335

Figure 4. Change of the general energy of system along coordinate of reaction R_{O7C22}.

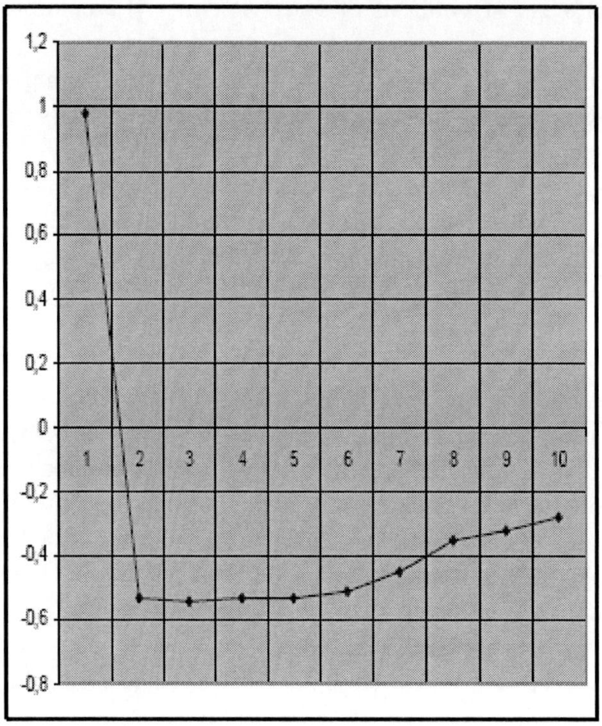

№ Step	R_{O7C22}, Å	q
1	2,8	0,98
2	2,6	-0,53
3	2,4	-0,54
4	2,2	-0,53
5	2,0	-0,53
6	1,8	-0,51
7	1,6	-0,45
8	1,4	-0,35
9	1,3	-0,32
10	1.2	-0,28

Figure 5. Change of a positive charge on atom of phosphorus P_{10} along coordinate of reaction R_{O7C22}.

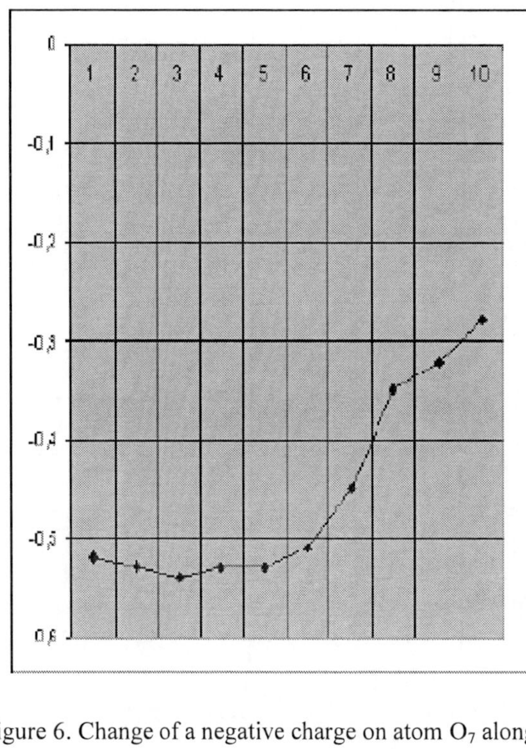

№ Step	R_{O7C22},Å	q
1	2,8	-0.52
2	2,6	-0.53
3	2,4	-0.54
4	2,2	-0.53
5	2,0	-0.53
6	1,8	-0.51
7	1.6	-0.45
8	1,4	-0.35
9	1,3	-0.32
10	1.2	-0.28

Figure 6. Change of a negative charge on atom O_7 along coordinate of reaction R_{O7C22}.

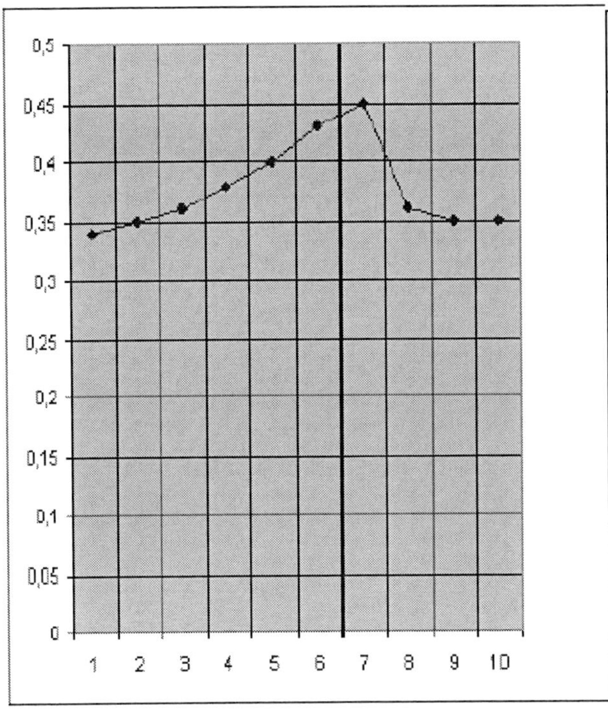

№ Step	R_{O7C22},Å	q
1	2,8	0.34
2	2,6	0.35
3	2,4	0.36
4	2,2	0.38
5	2,0	0.4
6	1,8	0.43
7	1,6	0.45
8	1,4	0.36
9	1,3	0.35
10	1.2	0.35

Figure 7. Change of a positive charge on atom of carbon C_{22} along coordinate of reaction R_{O7C22}.

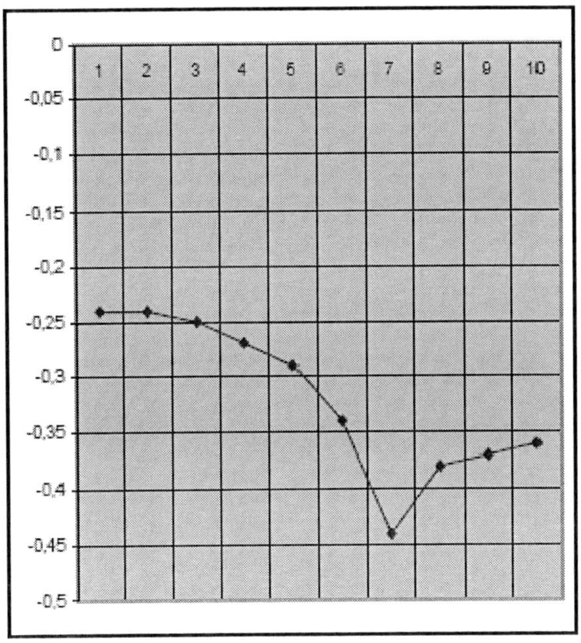

№ Step	R_{O7C22},Å	q
1	2,8	-0.24
2	2,6	-0.24
3	2,4	-0.25
4	2,2	-0.27
5	2,0	-0.29
6	1,8	-0.34
7	1,6	-0.44
8	1,4	-0.38
9	1,3	-0.37
10	1.2	-0.36

Figure 8. Change of a negative charge on atom Cl, along coordinate of reaction R_{O7C22}.

REFERENCES

[1] Clark T. *The Computer chemistry*. - M.: Mir. 1976. 512 p.

[2] Schmidt M.W., Baldrosge K.K., Elbert J.A.et. al. //*J. Computer Chem.* 1993. Vol. 14. P. 1347-1363.

[3] Bode B.M. and Gordon M.S. // *J. Mol. Graphics Mod.* 1998. Vol. 16. P. 133-138.

In: Chemical and Biochemical Kinetics: New Perspectives ISBN 978-1-61209-224-9
Editor: Gennady Zaikov

Chapter 5

ESTIMATION OF ACID FORCE OF COMPONENTS OF SYNTHESIS OF 5-ACETYLOXYMETHYL-2-CHLORO-5-ETHYL-1,2,3-DIOXAPHOSPHORYNANE

***V. A. Babkin,*[*1] *V. U. Dmitriev,*[1] *G. A. Savin,*[2,†] *E. S. Titova,*[3] *and G. E. Zaikov*[4,‡]**

[1]Volgograd StateArchitect-BuildUniversity Sebrykov Departament, Michailovka, Volgograd region, Russia
[2]Volgograd State Pedagogical University, Volgograd; Russia
[3]Volgograd State Technical University, Volgograd; Russia
[4]Institute of Biochemical Physics Russian Academy of Sciences, Moscow, Russia,

ABSTRACT

Quantum-chemicalcalculation of molecule of 5-acetyloxymethyl-2-chloro-5-ethyl-1,2,3-dioxaphosphorynane (I) and components of its synthesis: acetyl chloride (II) и 4-ethyl-2,6,7-trioxa-1-phosphabicyclo[2.2.2]octane (II) for the first time is executed by methodAB INITIO in basis 6-311G**. The optimized geometrical and electronic structures of these compounds are received. Their acid force is theorized. All these compounds concern to a class very weak C-H-acids (pKa > 14, pKa- the universal parameter of acidity). Dependence between acid force of components pKa (II) and pKa (III) of synthesis and acid force pKa (I) a required product is positioned:pKa (I) = pKa (II) - $\sqrt{pKa(III)+14.7}$

Keywords: quantum-chemical calculation, method AB INITIO, acidic force, 5-acetyloxymethyl-2-chloro-5-ethyl-1,2,3-dioxaphosphorynane, acetyl chloride, 4-ethyl-2,6,7-trioxa-1-phosphabicyclo[2.2.2] octane

* E-mail: sfi@reg.avtlg.ru.
† E-mail: gasavin@mail.ru.
‡ E-mail: chembio@sky.chph.ras.ru.

Aims and Backgrounds

5-Acetyloxymethyl-2-chloro-5-ethyl-1,2,3-dioxaphosphorynane(I) can be rather perspective start substance for reception of various chemical compounds, and in particular, the medical products possessing useful and probably unique properties.

This compound is received by interaction of acetyl chloride (III) and 4-ethyl-2,6,7-trioxa-1-phosphabicyclo[2.2.2] octane(II) in a gas phase:

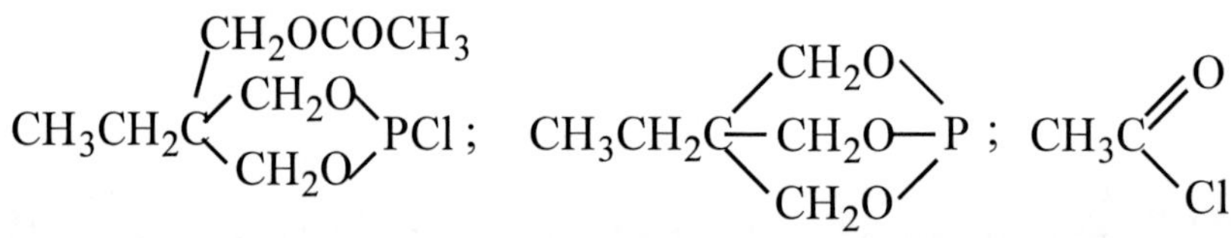

I **II** **III**

The mechanism of reaction of direct acylation of bicyclophosphites by chlorine anhydride of carboxylic acids now is not investigated. One of the first investigation phases of the mechanism of synthesis of studied compound (I) can consider an estimation of acid force of components of synthesis pKa (II) and pKa (III) and a required product pKa (I) and an establishment of dependence between them.

The Purpose of the present work was quantum-chemical calculation of components of synthesis of acetyl chloride (III) and 4-ethyl-2,6,7-trioxa-1-phosphabicyclo[2.2.2]octane(II) and a product 5-acetyloxymethyl-2-chloro-5-ethyl-1,2,3-dioxaphosphorynane (I) by method AB INITIO in basis 6-311G **, a theoretical estimation of their acid force and an establishment of dependence between acid force of components of synthesis pKa (III) and pKa (II) and pKa (I).

Methodical Part

Method AB INITIO in basis 6-311G ** with optimization of geometry on all parameters by the standard gradient method which has been built in in PC GAMESS, used for quantum-chemical calculationof components of synthesis of 5-acetyloxymethyl-2-chloro-5-ethyl-1,2,3-dioxaphosphorynane (I) [1]. Calculation was carried out in approach of the isolated molecule a gas phase. The theoretical estimation of acid force of components of synthesis was carried out under the formula [2]:

$$pKa = 49.04 - 134.61 \cdot q_{MAX}^{H+},$$

where pKa - a universal parameter of acidity, and q_{MAX}^{H+} - the maximal charge on atom of hydrogen of a molecule.

For visual representation of models of components of synthesis program MasMolPlt [3] was used.

RESULTS OF CALCULATIONS

The Optimized geometrical and electronic structures, the general energy, electronic energy, lengths of bonds and valent corners of components of synthesis ofacetyl chloride (III) and 4-ethyl-2,6,7-trioxa-1-phosphabicyclo[2.2.2]octane (II) and a product 5-acetyloxymethyl-2-chloro-5-ethyl-1,2,3-dioxaphosphorynane (I) are received by method AB INITIO in basis 6-311G ** and presented on Figures 1-3 and in Tables 1-4. Values pKa of all components of synthesis are certain by means of the formula:

$$pKa = 49.04-134.61 \cdot q_{MAX}^{H+}$$

$q_{MAX}^{H+}(I) = +0.16$, $q_{MAX}^{H+}(II) = +0.11$ and $q_{MAX}^{H+}(III) = +0.14$. Accordingly, pKa (I) = 27.5, pKa (II) = 34.2 and pKa (III) = 30.1. By us it has been positioned, that between acid force of components of synthesis pKa (II) and pKa (III) and acid force of a received product pKa (I) there is a following dependence:

Table 3. Optimized lengths of bonds, valency corners and charges of atoms of a molecule of 5-acetyloxymethyl-2-chloro-5-ethyl-1,2,3-dioxaphosphorynane

Bond lengths	R, Å	Valency corners	Degree	Atom	Charge (by Milliken)
C(1)-C(2)	1.53			C1	-0.23
C(2)-C(3)	1.55	C(3)C(2)C(1)	118	C2	-0.20
C(3)-C(4)	1.54	C(4)C(3)C(2)	104	C3	-0.34
C(4)-C(5)	1.54	C(5)C(3)C(2)	109	C4	0.15
C(5)-C(6)	1.53	C(6)C(3)C(2)	110	C5	0.13
C(4)-O(7)	1.41	O(7)C(4)C(3)	115	C6	0.12
C(5)-O(8)	1.42	O(8)C(5)C(3)	112	O7	-0.43
C(6)-O(9)	1.41	O(9)C(6)C(3)	112	O8	-0.68
O(8)-P(10)	1.61	P(10)O(8)C(5)	118	O9	-0.66
C(1)-H(11)	1.08	H(11)C(1)C(2)	112	P10	1.01
C(1)-H(12)	1.08	H(12)C(1)C(2)	107	H11	0.09
C(1)-H(13)	1.08	H(13)C(1)C(2)	107	H12	0.11
C(2)-H(14)	1.08	H(14)C(2)C(3)	108	H13	0.09
C(2)-H(15)	1.08	H(15)C(2)H(14)	105	H14	0.11
C(4)-H(16)	1.08	H(16)C(4)O(7)	105	H15	0.12
C(4)-H(17)	1.07	H(17)C(4)H(16)	108	H16	0.11
C(5)-H(18)	1.07	H(18)C(5)O(8)	105	H17	0.14
C(5)-H(19)	1.08	H(19)C(5)H(18)	108	H18	0.16
C(6)-H(20)	1.08	H(20)C(6)O(9)	108	H19	0.11
C(6)-H(21)	1.08	H(21)C(6)H(20)	107	H20	0.11
P(10)-Cl(22)	2.06	Cl(22)P(10)O(8)	99	H21	0.12
O(7)-C(23)	1.32	C(23)O(7)C(4)	118	Cl22	-0.29
C(23)-C(24)	1.50	C(24)C(23)O(7)	111	C23	0.49
C(24)-H(25)	1.08	H(25)C(24)C(23)	110	C24	-0.24
C(24)-H(26)	1.08	H(26)C(24)C(25)	110	H25	0.12
C(24)-H(27)	1.08	H(27)C(24)C(25)	107	H26	0.12
C(23)-O(28)	1.30	O(28)C(23)O(7)	123	H27	0.15
				O28	-0.48

Table 4. The general energy, energy of bonds, the maximal charge on atom of hydrogen, a universal parameter of acidity of components of synthesis of 5-acetyloxymethyl-2-chloro-5-ethyl-1,2,3-dioxaphosphorynane

№	The Component of synthesis	E_0, kJ/mol	E_{el}, kJ/mol	q_{MAX}^{H+}	pKa
1.	5-Acetyloxymethyl-2-chloro-5-ethyl-1,2,3-dioxaphosphorynane	-3700998	-7080777	+0.16	27.5
2.	4-Ethyl-2,6,7-trioxa-1-phosphabicyclo[2.2.2]octane	-2095991	-3994417	+0.11	34.2
3.	Acetyl chloride	-1604983	-1998646	+0.14	30.1

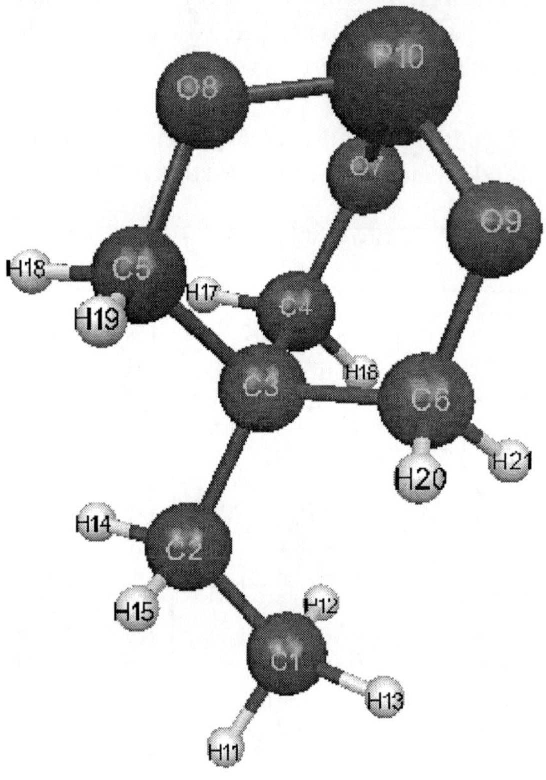

Figure 1. Geometric and electronic structure of a molecule of4-ethyl-2,6,7-trioxa-1-phosphabicyclo[2.2.2]octane(E_0=-2095991 kJ/mol, E_{el}=-3994417 kJ/mol).

$$pKa = pKa\ (II) - \sqrt{pKa(III) + 14.7}$$

The general energy (E_0), the electronic energy($E_{эл}$), the maximal charge on atom of hydrogen - $q_{MAX}^{H+}$and values pKa-a universal parameter of acidity of components of synthesis pKa (II) and pKa (III) and a received product pKa (I) are presented in Table 4.

Quantum-chemicalcalculationof a molecule5-acetyloxymethyl-2-chloro-5-ethyl-1,2,3-dioxaphosphorynane (I), and also components of its synthesis– acetyl chloride (III) and 4-

ethyl-2,6,7-trioxa-1-phosphabicyclo[2.2.2]octane(II) for the first time is executed by method AB INITIO in basis 6-311G **. The optimized geometrical and electronic structures of these compounds are received. Their acid force is theorized. By us it is shown, that all of them concern to a class weak C-H-acids (pKa> 14). Dependence between acid force of components of synthesis of acetyl chloridepKa (III) and 4-ethyl-2,6,7-trioxa-1-phosphabicyclo[2.2.2] octane (II) pKa (II), and also acid force of a received product 5-acetyloxymethyl-2-chloro-5-ethyl-1,2,3-dioxaphosphorynane (I) is positioned.

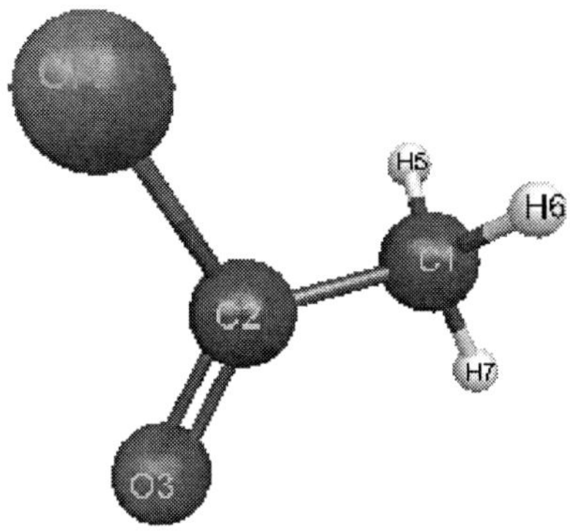

Figure 2. Geometrical and electronic structure of a molecule ofacetyl chloride(E_0=-1604983 kJ/mol, E_{el} =-1998646 kJ/mol).

Table 1. Optimized lengths of bonds, valency corners and charges of atoms of a molecule 4-ethyl-2,6,7-trioxa-1-phosphabicyclo[2.2.2]octane

Bond lengths	R, Å	Valency corners	Grad	Atom	Charge (by Milliken)
C(1)-C(2)	1.53			C1	-0.23
C(2)-C(3)	1.53	C(3)C(2)C(1)	116	C2	-0.18
C(3)-C(4)	1.53	C(4)C(3)C(2)	112	C3	-0.60
C(4)-C(5)	1.53	C(5)C(3)C(2)	108	C4	0.27
C(5)-C(6)	1.53	C(6)C(3)C(2)	112	C5	0.26
C(4)-O(7)	1.41	O(7)C(4)C(3)	110	C6	0.26
C(5)-O(8)	1.41	O(8)C(5)C(3)	111	O7	-0.71
C(6)-O(9)	1.41	O(9)C(6)C(3)	110	O8	-0.71
O(8)-P(10)	1.62	P(10)O(8)C(5)	118	O9	-0.71
C(1)-H(11)	1.08	H(11)C(1)C(2)	109	P10	1.21
C(1)-H(12)	1.08	H(12)C(1)C(2)	107	H11	0.10
C(1)-H(13)	1.08	H(13)C(1)C(2)	107	H12	0.09
C(2)-H(14)	1.08	H(14)C(2)C(3)	108	H13	0.09
C(2)-H(15)	1.08	H(15)C(2)H(14)	106	H14	0.10
C(4)-H(16)	1.08	H(16)C(4)O(7)	107	H15	0.10
C(4)-H(17)	1.08	H(17)C(4)H(16)	108	H16	0.11
C(5)-H(18)	1.08	H(18)C(5)O(8)	107	H17	0.10
C(5)-H(19)	1.08	H(19)C(5)H(18)	108	H18	0.10
C(6)-H(20)	1.08	H(20)C(6)O(9)	107	H19	0.11
C(6)-H(21)	1.08	H(21)C(6)H(20)	108	H20	0.10
				H21	0.11

Table 2. Optimized lengths of bonds, valency corners and charges of atoms of a molecule of acetyl chloride

Bond lengths	R, Å	Valency corners	Grad	Atom	Charge (by Milliken)
C(1)-C(2)	1.50			C1	-0.24
C(2)-O(3)	1.15	O(3)C(2)C(1)	127	C2	0.30
C(2)-Cl(4)	1.79	Cl(4)C(2)C(1)	112	O3	-0.33
C(1)-H(5)	1.08	H(5)C(1)C(2)	109	Cl4	-0.14
C(1)-H(6)	1.08	H(6)C(1)C(5)	108	H5	0.14
C(1)-H(7)	1.08	H(7)C(1)C(5)	110	H6	0.14
				H7	0.14

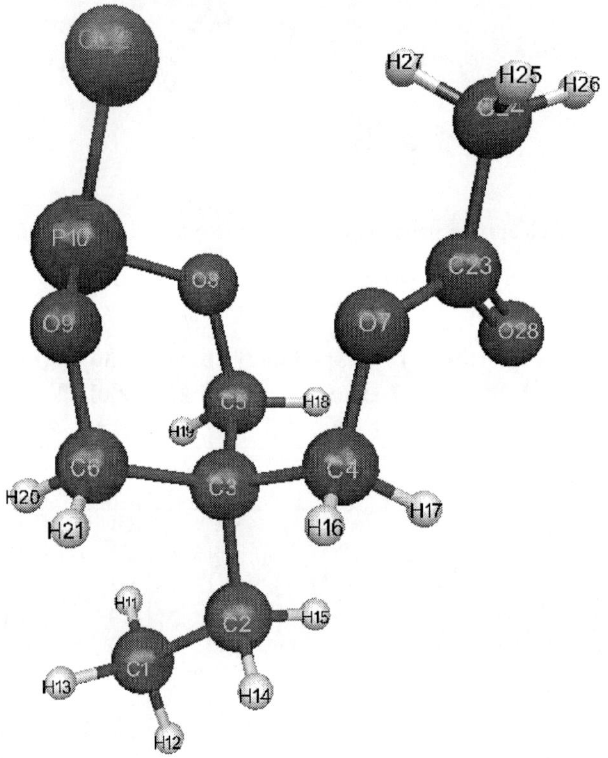

Figure 3. Geometrical and electronic structure of a molecule of5-acetyloxymethyl-2-chloro-5-ethyl-1,2,3-dioxaphosphorynane(E_0=-3700998 kJ/mol, E_{el}=-7080777 kJ/mol).

References

[1] Schmidt M.W., Baldrosge K.K., Elbert J.A.et. al. //*J. Computer Chem.* 1993. Vol. 14. P. 1347-1363.

[2] Babkin V.A., Fedunov R.G., Minsker K.S. et. al. // *Oxidationcommunication.* 2002. Vol. 25. № 1.P. 21-47.

[3] Bode B.M. and Gordon M.S. // *J. Mol. Graphics Mod.* 1998. Vol. 16. P. 133-138.

In: Chemical and Biochemical Kinetics: New Perspectives ISBN 978-1-61209-224-9
Editor: Gennady Zaikov

Chapter 6

QUANTUM–CHEMICAL CALCULATIONS OF ELECTRON STRUCTURE OF HEME AND ION OF IRON AND THEIR REACTIVITY IN REACTIONS WITH HYDROGEN PEROXIDE

A. A.Turovskyi,[1,*] V. I. Kopylets, L. I. Bazylyak,[2,†] A. R. Kytsya,[3,‡] and G. E. Zaikov[4,ä]

[1]Chemistry and Biotechnology of Combustible Minerals Division; Physical Chemistry of Combustible Minerals Department; Institute of Physical–Organic Chemistry and Coal Chemistry named after L. M. Lytvynenko; National Academy of Science of Ukraine, Lviv, Ukraine

[2]Physical–Chemical Methods of Metals Strengthening and Protection Division; Physical–Mechanical Institute named after G. V. Karpenko; National Academy of Science of Ukraine; Lviv, Ukraine

[3]Chemistry of Oxidizing Processes Division; Physical Chemistry of Combustible Minerals Department; Institute of Physical–Organic Chemistry and Coal Chemistry named after L. M. Lytvynenko; National Academy of Science of Ukraine, Lviv, Ukraine

[4]Kinetics of Chemical and Biological Processes Division; Institute of Biochemical Physics named after N. N. Emanuel; Russian Academy of Sciences, Moscow, Russia

ABSTRACT

It was done the quantum–chemical calculation of elementary reactions for heme of Iron and solvated ion of Iron with hydrogen peroxide interaction. It was obtained the

[*]E–mail: biotech@ukr.net.
[†]E–mail: kornii@ipm.lviv.ua.
[‡]e–mail: hop_vfh@ukr.net; bazyljak.L.I@nas.gov.ua.
[ä] e–mail: chembio@sky.chph.ras.ru.

potential surfaces for such reactions of interaction, heats and entropies of activation. It was shown, that the kinetic parameters$\Delta H^{\neq}$ and $\Delta S^{\neq}$ depend on ionization potentials (I) and electrons affinity (E) of the reactions reagents. Solvation of hydrogen peroxide by diphyllic agents influences both on electron characteristic of reagents (I, E) and on kinetic parameters of ion of Iron with hydrogen peroxide reactions. The Mikhaelis's complex of heme of Iron with hydrogen peroxide is thermodynamically nearer to the activated complex of reaction than the complex of Iron ion with hydrogen peroxide to the activated complex of its reaction. It was shown, that probably peptide chain of catalase attached to the heme of Iron can considerably influences on its electron characteristics which to great extentdetermine its reactivity.

Keywords: Quantum–chemical calculations, electron structure, heme of Iron

1. Introduction

For realization of catalytic function plenty of ferments contain into molecule the unalbuminous fragment – *cofactors*, which play in catalysis the role of prosthetic group or *coferment*. Prosthetic group attached to the albuminous part of a fragment covalently or with the participation of all series of weaker forces (hydrogen bonds, hydrophobic or ion–ionic interaction), which orientate such catalytically active group relatively polypeptide chain. Cofactor playing the catalytic function should be remained as a constant as a result of the catalytic reaction.

If the verily prosthetic group is used as the cofactor, then it realizes the all catalytic cycle and is attached to the same molecule of fragment.

We assume that such prosthetic group for catalase is heme of *Iron*, which plays the main catalytic function of hydrogen peroxide decomposition.

2. Aim and Background

The aim of the presented work was:

- to carry out the quantum–chemical calculations of electron structure of heme of *Iron* (catalase) and to clear its influence on catalase with the hydrogen peroxide interaction kinetic parameters;
- to compare the electron characteristics and reactivity in reactions of heme of *Iron* and ion of *Iron* with the hydrogen peroxide, since the primary reactions of such reagents are analogical.

3. Experimental Part

For carrying out of the all needed investigations it was used *the method of quantum–chemical calculation* of electron characteristics of reagents and potential surfaces.

Calculation of the reactions ways for heme of *Iron* and for ion of *Iron* interaction with the hydrogen peroxide was carried out with the use of semiempirical method *PM*6 using the *MOPAC*2009 programme. For the visualization of calculated structures the *«Jmol»* programme was used.

4. Results and Discussion

Letusnotethereactionofinteractionofhemeof*Iron*andionof*Iron*withthehydrogenperoxidevia K_1 and K_2, andthesamereactionswiththesolvatedmoleculeofhydrogenperoxide via K_1' and K_2' respectively.

For the calculations the following diphyllic molecules of solvates were as model:

- *glycol*$(HO–CH_2–CH(OH)–CH_3)$;
- *rhamnolipide*

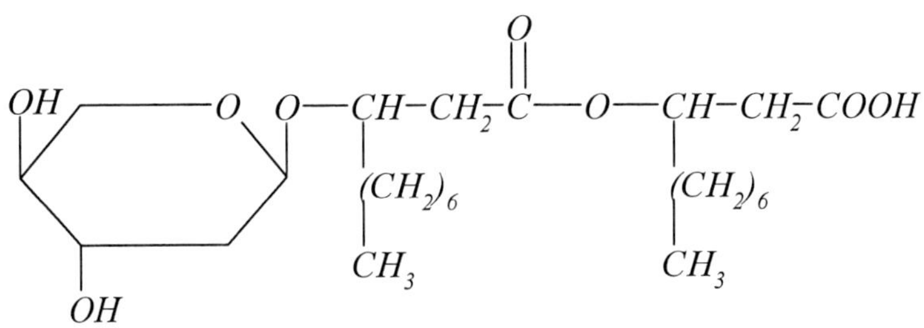

The optimal structure of reactive center of activated complex for reaction (K_1) of heme of *Iron* with the hydrogen peroxide is represented on *Figure* 1.

The distances *N–Fe* probably are hybrid as a result of the atom of *Iron* electrons delocalization and 1,9439; 1,8305; 1,8425; 1,9308 Å.

The optimal structure of the activated complex for the reaction of solvated by water of *Iron* ion with hydrogen peroxide is represented on Figure 2.

The every distance $(H_2O)...Fe^{+}(OH)$1,7533; 1,7625; 1,7684; 1,7450 Å. The distance *Fe...OH* is 1,5998 Å.

Practically, an interaction of the reagents into reactions K_1, K_2, K_1' and K_2' with the oxygen atom into hydrogen peroxide begins on average from the distance 2,8 Å. Into activated complexes in all variants the distances *Fe...O* in K_1, K_2, K_1' and K_2' are decreased till 1,91 Å, *i. e.* on 0,9 Å. The distance *O–O* in H_2O_2 consists of 1,21 Å into equilibrium state, and into activated complex this distance in all reactions on average is equal to 1,74 Å, *i.e.* is elongated on 0,5 Å.

Calculation of the potential surfaces of interaction reaction for heme and ion of *Iron* showed, that the activation heats of reactions consist of ~ 1,3 *kilocalories/mole* and 17 *kilocalories/mole* respectively (optimization gradient is 0,01).

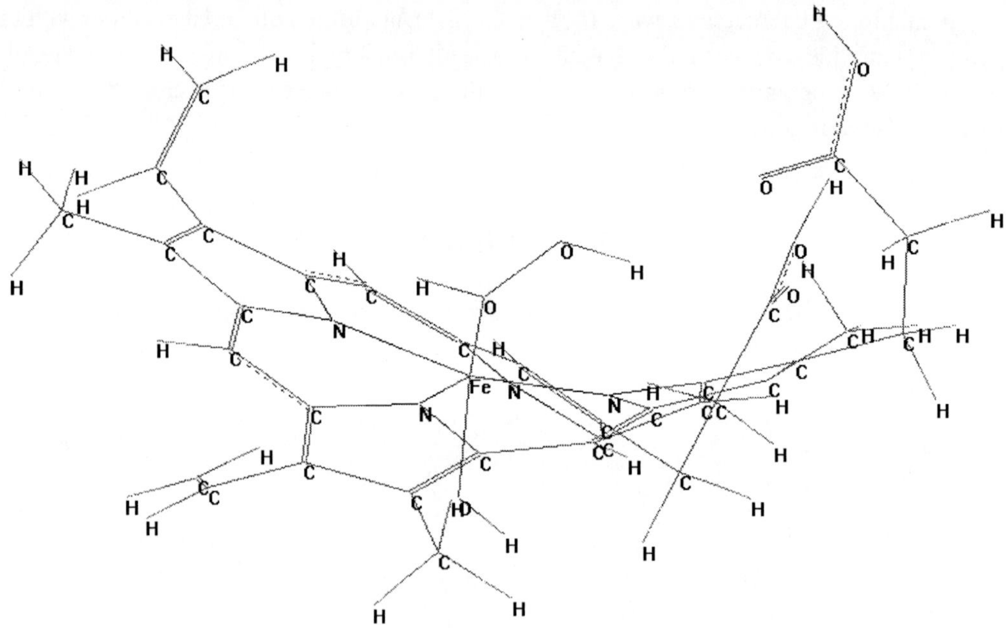

Figure 1. The optimal structure of reactive center of activated complex for reaction (K_1) of heme of *Iron* with the hydrogen peroxide.

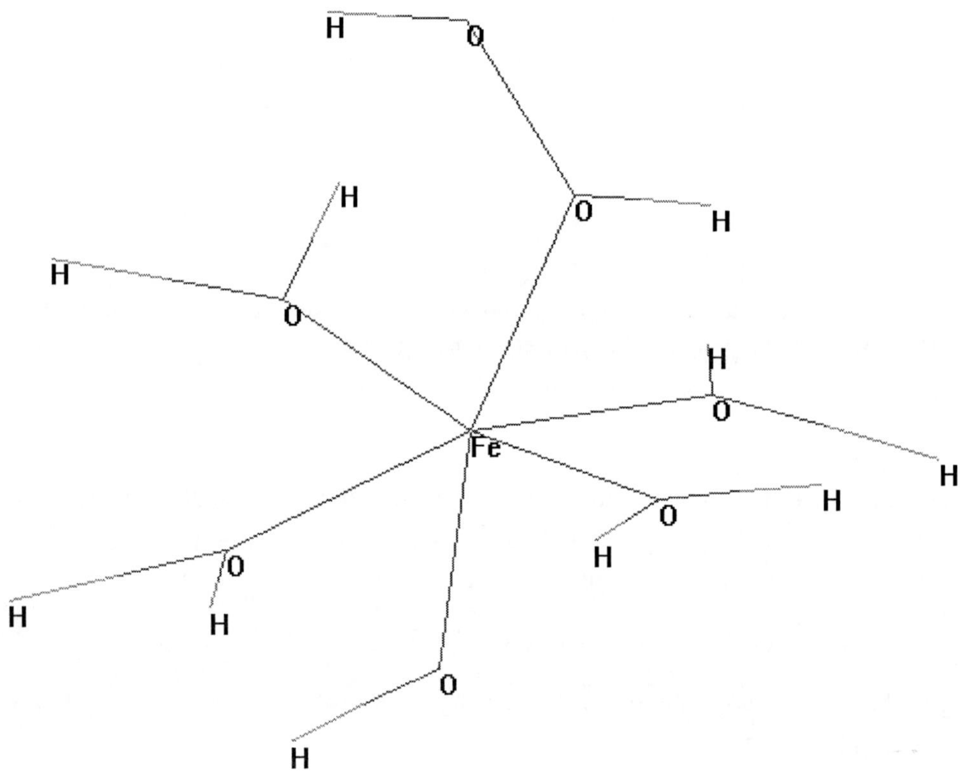

Figure 2.The optimal structure of the activated complex for the reaction of solvated by water of *Iron* ion with hydrogen peroxide.

Theheatsofactivationentropyfor the primary reactionsK_1, K_2, K_1' and K_2' are represented in Table 1.

The interesting is fact that the quantum–chemical calculations of kinetic parameters for reactions, considered in this work, in separate cases to great extent depend on the optimization gradient of system. The *Tables* 2–3 are illustrate this fact.

It can be seen from the *Table* 2, that the optimization gradient can cardinally influence on the calculation accuracy of activation heats for the reactions (variant K_1). However, into reactions K_2 such changes are less essential. Depending on the optimization gradient the reactions activation entropies are considerably changed.

Itisverycomplicated to explain the reasons of such phenomenon. Evidently, that the heme of *Iron* and the ion of *Iron* are differed by the quantity of the molecular orbitals. It is clear the fact that for the reaction of ion of *Iron* with hydrogen peroxide at great and little optimization gradients of reactive system we obtain approximately the same values of the activation heats, which can be explained by fact that the system hits at once on a top of the potential barrier, doesn't miss into some pseudominima. Whereas for reaction of heme of *Iron* with H_2O_2 it is observed very considerable difference into templates of activation depending on the optimization gradient, that, probably is caused by hit of the system into a range of the pseudominima of system energies that leads to the pseudo–activated complexes.

It can be seen from *Table* 3, that the values of the reactions activation entropies for interaction of reagents are considerably changed. For reactions K_1 and K_1' it is observed the great changes into $\Delta S^{\neq}$ depending on the optimization gradient. For K_2 and K_2' such changes are insignificant. Solvation of H_2O_2 leads to some $\Delta S^{\neq}$ increasing of reactions, that caused by decreasing of $\Delta S^{\neq}_{oscill.}$, *i. e.* the system with solvating reagent becomes more inflexible. The reasons of the reactions $\Delta S^{\neq}$ changes depending on optimization gradient are analogous to the calculations of the activation heats.

Table 1. The value $\Delta H^{\neq}$, $\Delta S^{\neq}$ and components of $\Delta S^{\neq}$ (optimization gradient is 0,01)

Reactions	$\Delta H^{\neq}$, kilocals/mole	$\Delta S^{\neq}$, general. osc/mole·degree	$\Delta S^{\neq}$, rect. osc/mole·degree	$\Delta S^{\neq}$, rotat. osc/mole·degree	$\Delta S^{\neq}$, oscill. osc/mole·degree
K_1	1,32	25,80	0,75	2,85	22,20
K_2	17,38	37,90	2,40	11,72	23,78
K_1'	1,157	22,63	2,16	4,12	16,35
K_2'	17,06	23,78	1,66	5,13	16,99

Table 2. The values of activations heats for the reactionsK_1, K_2, K_1' and K_2' depending on the optimization gradient

G, optimization gradient	$\Delta H^{\neq}$, kilocalories/mole			
	K_1	K_2	K_1'	K_2'
2,5	8,44	17,48	4,60	17,23
0,1	6,84	17,40	4,21	17,14
0,01	1,32	17,38	1,16	17,06

Experimental values of activation heats for ion of *Iron* and for heme of *Iron* of catalase interaction with the hydrogen peroxide consist of ~ 10*kilocalories/mole*and ~ 1 *kilocalories/*

mole respectively. ForthesolvatedmoleculeH_2O_2into reactionK_1'$\Delta H^{\neq}$< 1 *kilocalories/mole*. As we can see, theoretically calculated and experimentally founded values of the activation heats are agreed sufficiently satisfactory despite the fact that the tasks were solved via the semi–empirical approximate approach.

It can bee seen from the *Table* 1 that the activation heats for the interaction reaction of heme of *Iron* with H_2O_2 (K_1) are considerably less than the activation heats for the reaction of $Fe(OH)\cdot 5H_2O$ with H_2O_2 (K_2). For the solvated molecule of H_2O_2 the activation heat in K_1' is less than for the non–solvated one (K_1).However, the activation heats for the reaction of $Fe(OH)\cdot 5H_2O$ with H_2O_2 (K_2) in a cases of the non–solvated and solvated forms (K_2') are practically the same. The last it can be explained by the same type complexes formation for the reaction of $Fe(OH)\cdot 5H_2O\bullet H_2O_2$ and $Fe(OH)\cdot 5H_2O\bullet(H_2O_2)_{solvat.}$

From our point of view, the heats of activation for reactionsK_1 and K_2 are differed at the expense of different ionization potentials of reagents and reducing agents, *namely* heme of *Iron* and ion of *Iron* respectively. Really, it is shown from the *Table* 4, that the ionization potentials for K_1 and K_2 are enough different (the difference consists of 1,75 *eV* ~ 40 *kilocalories/mole*). The electron affinities are also differed into activated complex. Both first and second are favorable to great difference in values for activation heats for reactions K_1 and K_2. Ionization potentials for K_1' and $K_2^{'}$ into activated complex in reactions with solvated H_2O_2 are also so different (see *Table* 5). However, for K_2 and $K_2^{'}$ such difference is insignificant. The values of electron affinities for K_1 and K_2 are also essentially differed.

Table 3. The values of the reactions activation entropies ($\Delta S^{\neq}$) for reaction K_1, K_2, K_1' and $K_2^{'}$ depending on the optimization gradient

G, optimization gradient	ΔS$^{\neq}$, osc/mole·degree.			
	K_1	K_2	K_1'	$K_2^{'}$
2,5	12,6	36,2	9,32	27,4
0,1	20,4	37,40	13,68	28,3
0,01	27,8	37,90	22,63	23,78
	2,85 (rotat.)	11,72 (rotat.)	4,12 (rotat.)	5,13 (rotat.)
	2,75 (rect.)	2,40 (rect.)	2,16 (rect.)	1,66 (rect.)
	22,20 (oscill.)	23,78 (oscill.)	16,35 (oscill.)	16,99 (oscill.)

Table 4. The values of charges *(q)* on atoms*Fe* and *O* into reagents for the primary reactions

Reactions	Starting state		Activated complex		Finish state	
	Fe	*O* (H_2O_2)	*Fe*	*O* (H_2O_2)	*Fe*	*O* (H_2O_2)
K_1	+1,2635	–0,7428	+1,1214	–0,9833	+1,0855	–1,3124
K_2	+1,6350	–0,6933	+1,4217	–0,8854	+1,2063	–1,1655
K_1'	+1,2661	–0,7014	+1,1282	–0,9886	+1,1032	–1,3476
$K_2^{'}$	+1,6433	–0,6756	+1,5820	–0,9115	+1,2252	–1,1823

Table 5. The values of ionization potentials*(I)* for reagents in reactions for *Fe*–containing reagents and the values of electron affinity *(E)* in H_2O_2

Reactions	*I, eV,* for distance 2,8 $\overset{0}{A}$	*E, eV,* for distance 2,8 $\overset{0}{A}$	Activated complex	
			I, eV	*E, eV*
K_1	10,615 *(heme of Iron)*	6,435	9,083	5,19
K_2	12,362 *(ion of Iron)*	4,127	11,3010	3,22
K_1'	9,438 *(heme of Iron)*	6,822	8,7551	5,63
K_2'	12,055 *(ion of Iron)*	5,101	11,134	4,38

Let us note, that for the solvated molecule of H_2O_2 with rhamnolipid the electron affinity is something increased in comparison with the molecule of H_2O_2 solvated with the glycol and is considerably differed from *E* for non–solvated molecule of H_2O_2. It is necessary to notify also that the activation entropy (see *Table* 3) for the reaction K_1 is something higher than for reaction K_1' (solvated form); this is explained by some increasing the oscillation component of entropy of activated complex of reaction (*Table* 3), that is the activated complex for the reaction K_1 is more compact. The values of $\Delta S^{\neq}$ of reaction for K_1 and K_2 are sufficiently differed, mainly at the expense of $\Delta S^{\neq}$ rotation components. Sharply are differed the $\Delta S^{\neq}$ for K_2 and K_2' at the expense of $\Delta S^{\neq}$ *rotat.* and $\Delta S^{\neq}$ *oscill.* components. For the reactionsK_1 and K_2 a great difference of the activation heats is partially compensated by substantial difference of the activation entropies (especially are differed its oscillation components) (*Table* 3). Generally, the difference in $\Delta H^{\neq}$ for K_1 and K_2 can be explained by more thermodynamical expedience of *Mikhaelis's* complex (reaction K_1), in comparison with the complex $Fe(OH)\cdot 5H_2O\cdot H_2O_2$ (reaction K_2). The same situation is observed also in a case of reactions K_1' and K_2'. This will be shown in detail below.

Let us consider the separate electron characteristics of reagents into reactions of primary act of heme and ion of *Iron*. In both reactions K_1 and K_2 under primary elementary act we have in mind that the atom of *Iron* into heme and hydrated ion of *Iron* are donors of electrons, and the hydroxide peroxide is their acceptor, in other words we have the redox reaction.

The magnitudes of charges on the reactive centers of reagents into starting state and activated complex are represented in *Table* 4, from which it is seen, that the charges on the atom of *Iron* for reactionsK_1 and K_2 into equilibrium state are differed at the expense of electron–donor properties of substitutes. Into activated complexes the charges on *Fe* are decreased at the expense of growing the electron affinity in such complexes. In practically final products of reactions K_1 and K_2 the charges on *Fe* and *O* in H_2O_2 are some increased nearing to the charges of *Fe* into catalase. It is interesting to notify that the solvation effects have a little influence on the values of charges *Fe* in reactions K_1 and K_2 and also *O* in H_2O_2. Into activated complexes (*Table* 4) the charges on the atoms of *Fe* are decreased as a result of the electron affinity increasing of the reactive system. As we can see from *Table* 4, the charge on *Fe* in K_1 and K_2 for solvated and non–solvated forms of H_2O_2 via the reaction process is diminished, that is explained by transition of electron on the atom of reactive oxygenH_2O_2, that leads to the increasing of negative charge on the atom of O in H_2O_2.

Let us note that the spin on the atom of oxygen in H_2O_2 which doesn't react directly with the heme into activated complex is equal to 0,9615 and is near to the spin of radical $\overset{\bullet}{OH}$. For the reacting atom of oxygen in this state the spin is equal to 0,2534.

For the solvated molecule of H_2O_2 the spins into activated complex for the respective atoms of oxygen are equal to 0,9836 and 0,2855, that is in this case it is observed more localization of electrons on the respective atoms; this leads to some increasing of the electrons affinity and to the decreasing of the activation heats of elementary reactionsK_1' and K_2'. From our point of view very important electron characteristics of considered by us reactions K_1, K_2, $K_1^{'}$ and $K_2^{'}$ are *ionization potentials of reagents – reducing agents* (heme of *Iron*, ion of *Iron*) and *H_2O_2oxidizing agentelectron affinity* (*Table* 5). The values of ionization potentials for heme of *Iron* and solvated ion of *Iron* into equilibrium state consist of 10,631 and 12,354 *eV* respectively. When the reagents interaction process begins on distance 2,8 $\overset{0}{A}$ the ionization potentials are some changed (see *Table* 5). Greater changes of *(I)* are observed into activated complex. Solvation of hydroxide peroxide considerably changes the values of the electron affinity H_2O_2 into reactions $K_1^{'}$ and $K_2^{'}$ both for starting systems and for activated complex. Start of the interaction (the distance between atoms of reagents *Fe…O* consists of 2,8 $\overset{0}{A}$) greatly changes the electron affinity into the all considered reactions. It is observed also the substantial diminution of the electrons affinity for H_2O_2 for all reactions and for activated complex. It is seen from the *Table* 5, that the ionization potentials in reactions K_1 and K_2 are considerably differed that is logicallyreflected also on the values of the activation heats of these reactions. Some lower are ionization potentials of solvated reagents in reactions $K_1^{'}$ and $K_2^{'}$.

Electron affinity for solvated molecule H_2O_2 with diphyllic molecules for starting state is increased.Electron affinities into activated complex for K_1 and K_2 are enough different.Both ionization potentials and electrons affinities are considerably reflected on kinetic characteristics of reactionsK_1 and K_2, $K_1^{'}$ and $K_2^{'}$. It is necessary to notify, that the ionization potentials of heme of *Iron* and ion of *Iron* both for reagents at the beginning of interaction (2,8 $\overset{0}{A}$), and into the activated complex have an influence on kinetic parameters of reactions K_1 and K_2, $K_1^{'}$ and $K_2^{'}$.

If to assume, that the activation heats of considered reactionsK_1 and K_2, $K_1^{'}$ and $K_2^{'}$ depend both on the reagent ionization potential of electron donor and on the electron affinity (acceptor of the electrons) and to accept that the orbital energy can be estimated in accordance with the equation $\frac{const}{I-E}$, then we will obtain the ratio of orbital energies for the reactions K_1 and K_2 taking into account that the ionization potential and electrons affinities is equal to $K_2/K_1 = 2{,}07$. Respectively, the ratio of activation for such reactions is equal to 13,2.

For reactions$K_2^{'}$ and $K_1^{'}$ the ratio of the orbital energies is equal to 2,23 and the ratio of the activation heats in this case is equal to 14,7. Letusnote that the ratio of the activation heats of these reactions is proportional to the values of the orbital energies. Atthecalculated values of the ionization potentials and electron affinities at known values (which are also calculated), it can be estimated the kinetic parameters both for chemical and biological reactions.

It is interesting the variant of the solvated heme of *Iron* with the molecule of rhamnolipid. This case can be considered as the model influence of a link of peptide chainwhich is connected with the atom of *Iron* via carbonyl group on electron characteristics of heme.

Calculation of a dependence of enthalpy on rotation angle of rhamnolipid around the bond *Fe...O≡C<*. showed, that the rotation barrier is observed within the range 80^0 and consists of 3,5 *kilocalories/mole*.

In *Table* 6 there is a change of the ionization potentials of heme depending on rotation angle around the bond *Fe...O≡C<*.

For rotation angle of molecule within the range of $0–80^0$ the ionization potential of heme is increased on 0,364 *eV* (see *Table.* 6); this consists of ~8,4 *kilocalories/mole*. Within the range of $90–160^0$ the ionization potential is decreased on 0,378 *eV* (see *Table* 6), that consists of ~8,64 *kilocalories/mole*. Under the following rotation of rhamnolopid's molecule around the bond *Fe...O=O*, the ionization potential is more decreased and becomes lesser – as same as in the heme of *Iron* without rhamnolipid.

So, the difference of ionization potentials of substituted and unsubstituted hemes consists of 0,277 *eV* (~6,4 *kilocalories/mole*) (*Table* 6). Under the first approximation it can be assumed, that the peptide chain (especially the last link) can influences on electron structure of heme of *Iron*; that is, the ionization potential of heme is changed to the minimal value that in some manner can influences on the heat barrier decreasing of activation reaction for heme with the hydrogen peroxide.

If to assume, that the last link of peptide chain is rotated with not great barrier (in the last case it consists of ~3,5 *kilocalories/mole*) and, taking into account that the semi–empirical method gives the calculations of energy with enough great error (especially for the last not great peptide fragment), the rotation of link will be almost free. Therefore, the ionization potential of a system also will be minimal at the greatest rotation angle of link that leads to the catalysis of hydrogen peroxide molecule.

Table 6. Dependence of the ionization potential *(I, eV)* of heme on rotation angle *(θ, degree)*

I, eV	10,892	10,896	10,932	10,945	10,984	11,206	11,234	11,256	11,243
θ, degree	0	10	20	30	40	50	60	80	90
I, eV	11,105	10,987	10,962	10,902	10,881	10,880	10,878	—	—
θ, degree	100	110	120	130	140	150	160	—	—

Under the less rotation angles this catalysis reaction can proceeds with some less rates as a result of some increasing the ionization potential of heme; this is favorable to the formation of more thermodynamically advantageous *Mikhaelis's* complex and hence to the advantageous geometry of the activated complex formation for the reaction of heme with hydrogen peroxide.

CONCLUSIONS

From the all above–said it can be done the following conclusions.

Firstly, the primary reactions of the fermentative catalysis of heme of *Iron* with H_2O_2 and the reaction of ion of *Iron* proceed with the activation heats, which are sharply differed due to different ionization potentials of heme and ion of *Iron.*

Secondly, the solvation of H_2O_2 by diphyllic agents increases the values of the electron affinity of substrateH_2O_2 to electron and in such a way leads also to the decreasing of the activation heats and entropies for reactions of heme of *Iron* and ion of *Iron* with hydrogen peroxide.

Thirdly, the solvation of heme of *Iron* by diphyllic agents (rhamnolopids and *et al*) leads to the variation of ionization potential of system as a result of diphyllic reagent's molecule rotation around the bond $Fe...X$, that marks on reactivity of heme.

Fourthly, it can be assumed that the peptide radical of catalase can in some manner influences on the ionization potential of heme of *Iron.* Probably, such influence is essential only for links directly connected with the heme either chemically or coordinated. However, generalizing the above–said, it can be theoretically assumed, that the higher rates and the less heats and activations of fermentative reactions, in comparison with the analogical chemical ones, caused by different electron structures of ferments and chemical catalyst. The important electron characteristic of these catalysts is the ionization potential. Activation heat of reaction is proportional to the magnitude of this characteristic. For the trustworthiness of this conclusion it is necessary to carry out a series of experimental and theoretical investigations.

The important in reactions of fermentative catalysis is fact, that the *Mikhaelis's* complex geometrically and thermodynamically is enough neared to the activated complex.

Heme of *Iron* with the hydrogen peroxide forms the complex, which is transformed into activated complex *(AC)*:

$$heme\text{–}Fe\cdot H_2O_2 \rightarrow heme\text{–}Fe...H_2O_2(AC)$$

$$\Delta F heme\cdot Fe\cdot H_2O_2 = -268 \textit{ kilocalories/mole}$$

$$\Delta F^{\neq} \approx -7 \textit{ kilocalories/mole}$$

Then, a free energy of the activated complex is equal to –275 *kilocalories/mole.* In other words, the free energies of activated complex and *Mikhaelis's* complex are differed in 1,02 times. For the reaction of ion of *Iron*$Fe(OH)\cdot 5H_2O$ with H_2O_2 a free energy of complex $[Fe(OH)\cdot 5H_2O\cdot H_2O_2]$ is equal to 0,7 *kilocalories/mole.* Free energy of activated complex $[Fe(OH)\cdot 5H_2O...H_2O_2]$ consists of ~ 6,7 *kilocalories/mole.* That is, ΔF of activated complex is differed from ΔF of ion of *Iron* with H_2O_2 approximately *in 10 (!) times*, that greatly reflects on reactivity of ion of *Iron* with hydrogen peroxide.

So, transformation *of Mikhaelis's* complex into activated complex of fermentative reaction needs considerably less energy, than in a case of analogical chemical reaction.

In: Chemical and Biochemical Kinetics: New Perspectives
Editor: Gennady Zaikov
ISBN 978-1-61209-224-9

Chapter 7

STRUCTURE AND PROPERTIES OF POLYMER ADSORPTION LAYERS ON INORGANIC PIGMENTS SURFACE IN AQUEOUS DISPERSION FORMED UNDER ULTRASONIC TREATMENT

N. Bulychev,[1,2] E. Kisterev,[1] K. Dirnberger, V. Zubov,[3] and C. D. Eisenbach[2,4]

[1]N.S. Kurnakov Institute of General and Inorganic Chemistry of Russian Academy of Sciences, Moscow, Russia
[2]Institute forPolymer Chemistry, University of Stuttgart, Stuttgart, Germany
[3]Lomonosov Moscow State Academy of Fine Chemical Technology, Moscow, Russia
[4]Research Institute for Pigments and Coatings, Stuttgart, Germany

ABSTRACT

The course and the result of the surface modification of titanium dioxide and ferrous oxide pigments in aqueous dispersion by ethylhydroxyethylcellulose (EHEC) without and with mechanical treatment of the dispersion by ultrasonic power was studied by the electrokinetic sonic amplitude (ESA) method. The evaluation of the ESA data showed that the ultrasonic treatment causes a significant thickness increase of the EHEC layer on the pigment which is primarily attributed to the ultrasonically induced activation of the pigment surface. The thickness of the polymer adsorption layer derived from ESA measurements was confirmed by TEM investigations. The ultrasonic treatment leads to significant changes of the properties of the adsorption layers and is a promising method for the surface modification of pigments with regard to dispersion stability.

Keywords: aqueous dispersions, polymers, electrokinetic sonic amplitude method, polymer adsorption layers, ultrasonic treatment

AIM AND BACKGROUND

Recent developments in research of dispersed systems have shown that mechanical, in particular ultrasonic treatment of aqueous dispersions of pigments in the presence of polymeric stabilizers leads to a significant enhancement of the stability of these dispersions in comparison to dispersions prepared without ultrasonic treatment. However, the effect of the ultrasonic treatment on the pigment-polymer interaction and structure formation of adsorption layers, especially for novel tailor-made amphiphilic copolymers, is still obscure and needs further investigation for elucidation of the phenomenaobserved.The main aim of this paper is to get more insights on the effects of the ultrasonic treatment on the polymer solid-liquid interface behaviour and the formation of polymer adsorption layers.

INTRODUCTION

In recent years, aqueous colloidal dispersions of pigments have been of increasing interest from both scientific and practical points of view. They are important, ecologically friendly colloidal systems widely used in polygraphic and paint industries. The pigment particles are usually of 200 – 2000 μm diameter and strongly aggregated. In order to achieve a good stabilization in aqueous pigment dispersions, many formulations were proposed [1-8].Earlier we reported about the role of mechanical, e.g. ultrasonic treatment for obtaining of highly stable dispersions [7,9]. It was shown, that using of polymer surfactants in combination with mechanical action can significantly improve the quality of dispersed systems. However, specific aspects of the pigment-polymer interaction and structure formation of adsorption layers especially under mechanical, i.e., ultrasonic action have not yet been studied in detail.

Electrokinetic sonic amplitude (ESA) measurements have been demonstrated to be a powerful method that could provide the desired information about the process of polymer adsorption [7-9] and was employed in this work. Commercial ethylhydroxyethylcellulose (EHEC) was used as stabilizer of TiO_2 and Fe_2O_3 aqueous dispersions. The polymer layers created on the particle surfaces in the absence and presence of mechanical (ultrasonic) action were investigated.

EXPERIMENTAL PART

Materials

As pigments, the titanium dioxide rutil pigment Kronos 2310 with particle size 0.3 μ, ferrous oxide Bayerfox with particle size 0.3 μ were employed as received. EHEC with molecular weight 60.000 was provided by Berol.

Techniques

For the preparation of aqueous pigment dispersions, the pigment was added to water or to the aqueous polymer solution and dispersing of the pigment was first achieved by means of a laboratory stirrer (700 rpm for 10 min). When ultrasonification was applied, the system was subsequently treated with ultrasound for 2 min with an ultrasonic generator UZDN-2 or Branson Sonifier B-12 with actual power of 1.5 W/cm^2.

The pigment-polymer interaction and the polymer adsorption layer formation were investigated by electrokinetic sonic amplitude(ESA) measurements as described elsewhere [12]. The particle size was measured by both ESA (cf. [12]) and by using Nanosizer (Malvern Instruments). Standard optical equipmentwas used for refractive indexmeasurements. Transmission electron microscopy (TEM) investigations were carried out according to the standard procedure. The samples of the pigment dispersions were diluted with distilled water by a factor of 100, then dropped on the TEM grid of 3 mm diameter and dried on air for 1 day. The TEM pictures were taken by using a Jeol 2000 FX device under the voltage of 200 kV.

Results and Discussion

Data reported in literature [7,9,10], showed that mechanical, in particular, ultrasonic treatment of aqueous dispersions of pigments in the presence of polymeric stabilizers leads to a significant enhancement of the stability of these dispersions as compared to dispersions prepared without ultrasonic treatment. It was proven by IR-analysis that the thickness and stability of the polymer adsorption layers were increased and improved by the ultrasonification [9]. The optimal parameters of the ultrasonic treatment and their influence on the properties of dispersed systems were established [11]. However, the effect of the ultrasonic treatment on the pigment-polymer interaction and polymer adsorption is still obscure and needs further investigation for better elucidation of the phenomena observed.

ESA measurements of aqueous dispersions of TiO_2, stabilized by ethylhydroxyethyl-cellulose (EHEC), give quantitative information about the process of adsorption of polymers as reflected from first the dependence of the dynamic mobility μ on the relative polymer concentration (cf. [12-14]) as shown in Figure 1 and 2, and second from the comparison of this dependency for systems without and with ultrasonic treatment (Figure 2).

As one can see from the comparison of Figure 1 and 2, ultrasonic treatment has a distinct effect on the behavior of the particulate pigment-polymer suspension. Without ultrasonic action, the formation of a polymeric adsorption layer on the pigment surface seems to be reached at 1 wt.-% EHEC in relation to the pigment concentration (saturation concentration as indicated by the unchanged dynamic mobility with further increase of polymer concentration) followed by polymolecular adsorption between about 2 and 3 %. In the presence of ultrasonic action, first the initial dynamic mobility of the pristine TiO_2 is much higher, and the saturation concentration of polymer is only reached at about 15 %. Further addition of polymer does not effect the dynamic mobility.

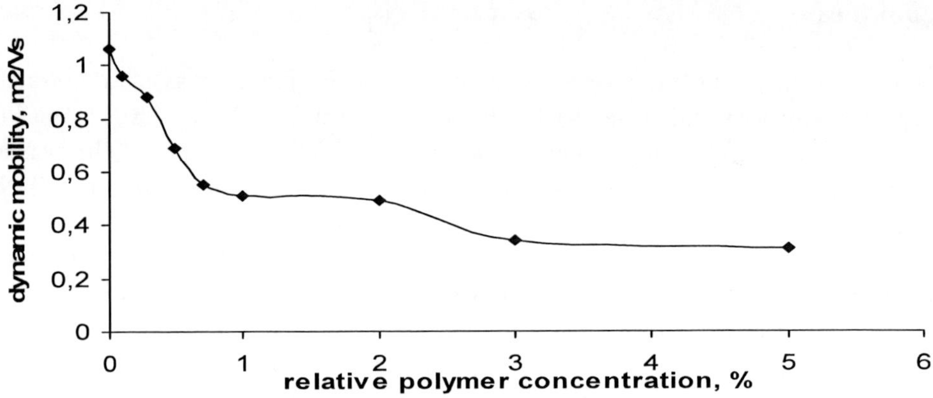

Figure 1. Dependence of the dynamic mobility on the relative concentration of EHEC for 1 wt.-% TiO_2 aqueous dispersion without ultrasonic treatment.

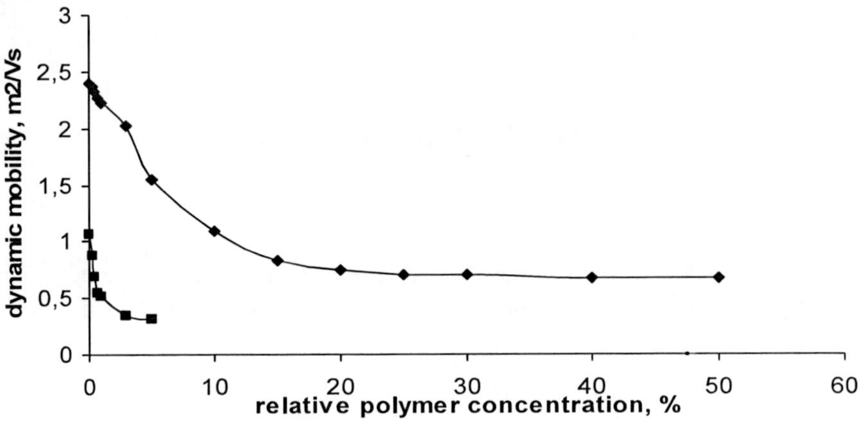

Figure 2. Dependence of the dynamic mobility on the relative concentration of EHEC for 1 wt.-% TiO_2 aqueous dispersion with (◆) and without (■, cf. Figure 1) ultrasonic treatment.

However, e.g., for the system with 50 wt.-% polymer relative to the 1 wt.-% pigment dispersion, it was observed that the dynamic mobility was further decreased to 0.35 m^2/Vs after 1 day. This means that even though the ultrasonic power was shut off, polymer deposition still occurred, indicating that polymolecular adsorption in case of treated samples proceeds over a longer period of time. Obviously the ultrasonic action not only increases the pigment surface by creating a more fine dispersion but also activates the pigment surface leading to an ultimately higher polymer adsorption in comparison to the non-treated samples. Similar results were obtained for Fe_2O_3 aqueous dispersions as revealed from Figure 3, 4.

Comparing the values of saturation concentration without and after ultrasonic treatment, one can infer that the amount of polymer adsorbed on the particle surface significantly increases when ultrasonification is applied. In this context it has to be considered that the ultrasonic treatment leads to a finer dispersion by decreasing the particle size, and, consequently leading to an increase of the surface area prone to polymer deposition. Thus both the increase in the total surface area of the dispersed particles and a possible ultrasonically induced activation of the particle surface must be considered as being responsible for the observed effects.

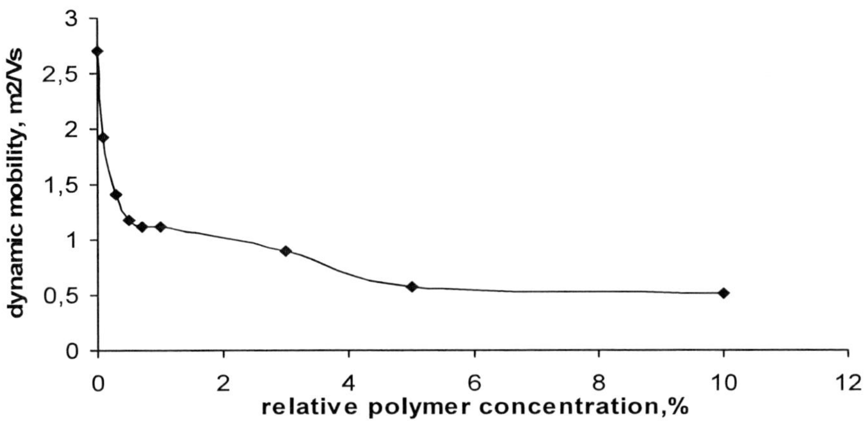

Figure 3. Dependence of the dynamic mobility on the relative concentration of EHECfor 1 wt.-% Fe_2O_3 aqueous dispersion without ultrasonic treatment.

To get some quantitative information about the particulars of the polymer adsorption and to explain the increased amount of polymer adsorbed when ultrasonic treatment was applied, the surface area of the uncoated pigment and of the polymer coated pigment were calculated; these data were compared with information obtained from TEM pictures of the particulate system and further related to data derived from ESA measurements.

First, the total surface area of the particles as related to the particle radius was calculated:

The total surface area S_{tot} of all particles in the dispersion can be expressed by:

$$S_{tot} = S_{part} * n_{part} \tag{1}$$

Here S_{part} is the surface area of a single particle, and n_{part} is the number of particles.

Assuming the particles to be spherical, the surface area of one particle can be calculated from the particle radius r_{part}:

$$S_{part} = 4 * \pi * r^2_{part} \tag{2}$$

The number of particles in the dispersion as related to the particle radius is given by:

$$n_{part} = \frac{m_{total}}{m_{part}} = \frac{m_{total}}{V_{part} * \rho_{TiO_2}} = \frac{3 * m_{total}}{4 * \pi * r^3_{part} * \rho_{TiO_2}} \tag{3}$$

Here m_{total} is the total mass of the pigment in the dispersion, m_{part} is the mass of a single particle, V_{part} is the single particle volume, and ρ_{TiO2} is the density of titanium dioxide.

Thus, the formula for the total surface area of all particles in the dispersion is expressed by:

$$S_{tot} = \frac{4*\pi*r^2{}_{part}*3*m_{total}}{4*\pi*r^3{}_{part}*\rho_{TiO_2}} = \frac{3*m_{total}}{r_{part}*\rho_{TiO_2}} \tag{4}$$

The ratio of the total particle surface area of the ultrasonically treated (index 2) and non-treated (index 1) dispersions is given by:

$$\frac{S_{tot_2}}{S_{tot_1}} = \frac{3*m_{total}*r_{part_1}*\rho_{TiO_2}}{3*m_{total}*r_{part_2}*\rho_{TiO_2}} = \frac{r_{part_1}}{r_{part_2}} = \frac{d_{part_1}}{d_{part_2}} \tag{5}$$

Thus, the ratio between the values of the particle surface area $S_{tot\,2}/S_{tot\,1}$ can be expressed as the reverse ratio of the average particle diameters, i.e., $\left(d_{part\,2}/d_{part\,1}\right)^{-1}$.

Using this equation and by inserting the values of the particle size at the saturation concentration of ultrasonically treated and non-treated samples of TiO_2 and Fe_2O_3 dispersion aqueous in EHEC-solution, it is possible to calculate the increasing of particles surface area during the ultrasonic treatment. These data are given in Table 1.

From Table 1 it is seen that for both systems the particle surface area increases by a factor of 3.4 – 4.7 upon ultrasonic treatment of the dispersion. However, the saturation concentrations for these systems increase by a factor of 7 – 15. This indicates that the amount of polymer adsorbed per unit of the particle surface after ultrasonic treatment is higher in comparison with the non-treated samples; this also means an increased thickness of the adsorption layer in the case of ultrasonically treated dispersions.

The difference of the dynamic mobility $\Delta\mu$ between the pure and EHEC coated pigments (at the saturation concentration) is higher for the ultrasonically treated ($\Delta\mu_2$) than the non-treated ($\Delta\mu_1$) samples (see Table 2). The proportionality between the $\Delta\mu_2/\Delta\mu_1$ ratio (last column Table 2) of the two pigment systems is similar as the ratio SC_2/SC_1 of the saturation concentration (second column Table 2).

Thus the correlation of the ESA measurement data with regard to the established saturation concentration and change in the dynamic mobility gives a conclusive picture of the effects of ultrasonic treatment on the pigment/EHEC interaction which is in accordance with the calculations of the particle surface/particle diameter relation (Eq. 5).

As it was proven by refraction index measurements of the supernatant liquid of sedimented pigment dispersions, up to the saturation concentration of the polymer, all polymer added to the system is adsorbed on the particle surface; only when this concentration was exceeded, a change in the index of refraction due to dissolved polymer was observed. Thus, since the amount of polymer adsorbed and the surface area of the particles is known, the thickness of adsorption layers is given by Eq. (6) where the total area of the particles S_{tot} is calculated by the equation (4):

$$d = \frac{V_{tot}}{S_{tot}} \tag{6}$$

The volume of the adsorption layer V_{tot} is given by the ratio of the mass of polymer adsorbed $m_{tot\ pol}$ to the density ρ_{pol} of the polymer (Eq. 7):

$$V_{tot} = \frac{m_{tot\ pol}}{\rho_{pol}} \tag{7}$$

The relationships of Eq. (6) and (7) allow to calculate the thickness of the adsorption layers on the pigment surface for both the ultrasonically treated and untreated pigment dispersions. The data are presented in Table 3. One can see that the thickness of the adsorption layer increases by the ultrasonic treatment for both systems tested. This confirms that during the ultrasonic treatment an activation of the pigment surface occurs.

Aqueous dispersions of TiO_2 and Fe_2O_3 stabilized by EHEC were also investigated by transmission electron microscopy (TEM) in order to visualize the change of the dispersed system upon ultrasonic treatment, and also to confirm the above conclusions about the thicknesses of the EHEC polymer adsorption layer.

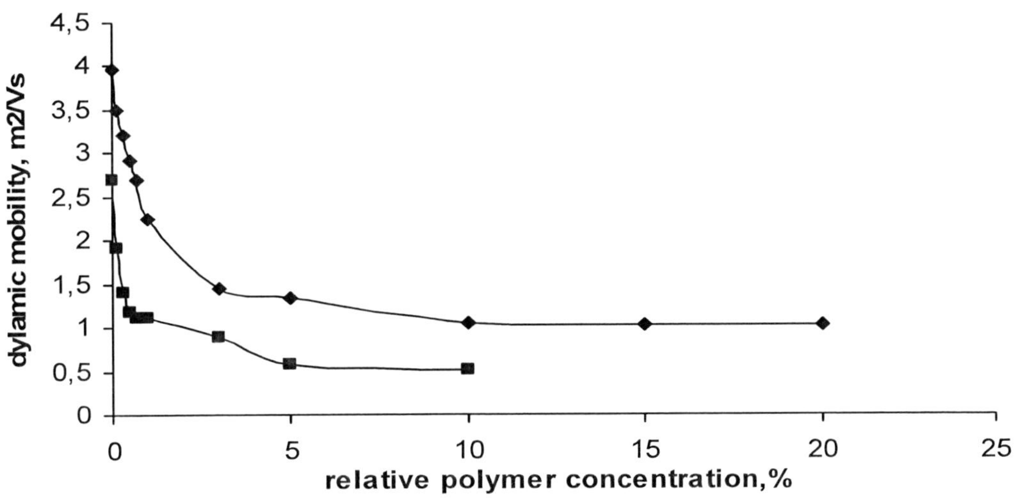

Figure 4. Dependence of dynamic mobility on relative concentration of EHEC for 1 wt.-% Fe_2O_3 aqueous dispersion with (◆) and without (■, cf. Figure 1) ultrasonic treatment.

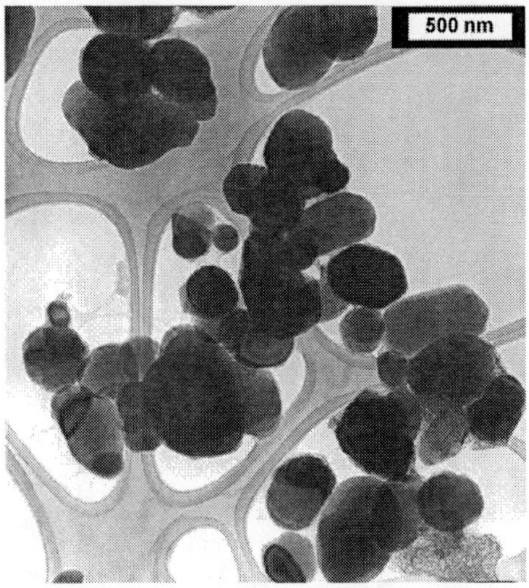

Figure 5. TEM picture of TiO_2 aqueous dispersion without polymer stabilizer and without mechanical (ultrasonic) treatment.

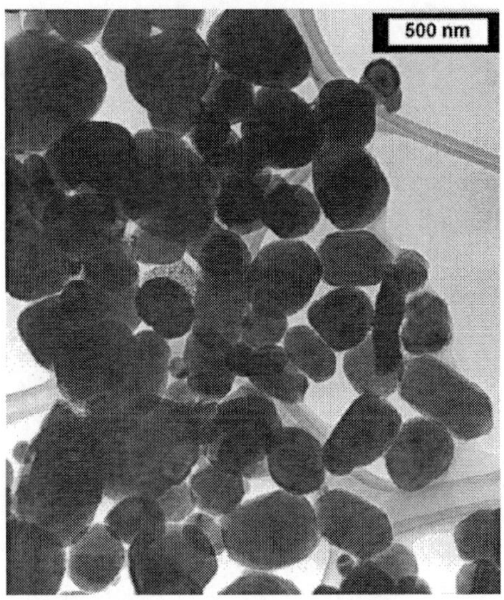

Figure 6. TEM picture of TiO_2 aqueous dispersion without polymer stabilizer after ultrasonic treatment.

From the TEM picture Figure 5 it is seen that the dispersion of the TiO_2 without added polymer stabilizer and without ultrasonic treatment consists of aggregates of about 5 – 10 particles; the size of such aggregates of the non-treated system is about 1 μm and is in agreement with Nanosizer and ESA data. However, the dispersion of the TiO_2, again without EHEC, after ultrasonic treatment consists of extended aggregates (Figure 6) of sizes up to 4 – 5 μm; this size increase is due to the activation of the pigment particle surface by the ultrasonic power. As a result of this coagulation, fast sedimentation of the dispersion occurs. From Figure 5 and 6 it is also evident that the primary particles are spherical in shape; this confirms the assumption made above in the calculation of the particle surface area.

When the dispersion of the TiO_2 or Fe_2O_3 pigments is carried out with added EHEC, the polymer acts as a stabilizer and primary particle dispersions are obtained. As it is seen from the comparison of the TEM pictures of the TiO_2 or Fe_2O_3 aqueous dispersions stabilized by EHEC (polymer concentration chosen corresponding to the saturation concentration as obtained from ESA measurements, Figure 1 – 4) which were obtained without (Figure 7 and 9) and with ultrasonic treatment (Figure 8 and 10), the ultrasonification leads to an increased thickness of the EHEC coating layer around the primary particles.

These findings are in agreement with the much higher polymer saturation concentration required in the ultrasonically treated dispersions (see Figure 2 and 4) as compared to the non-treated systems (see Figure 1 and 3) which has already been discussed above (Table 1 and Table 2). The thickness of the adsorption layers strongly increases from 4 – 6 μm of the not ultrasonically treated dispersions to about 15 – 20 μm when ultrasonic treatment was applied. This is in rough agreement with the values calculated above from ESA data (see Table 3).

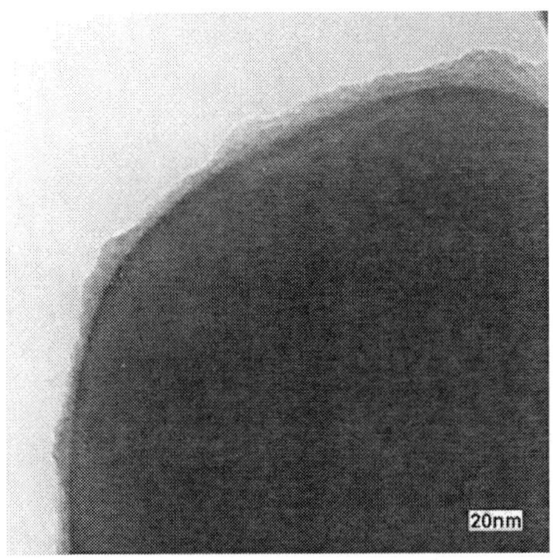

Figure 7. TEM picture of TiO_2 primary particle aqueous dispersion stabilized by EHEC (1 wt.-% relative pigment) without mechanical (ultrasonic) treatment.

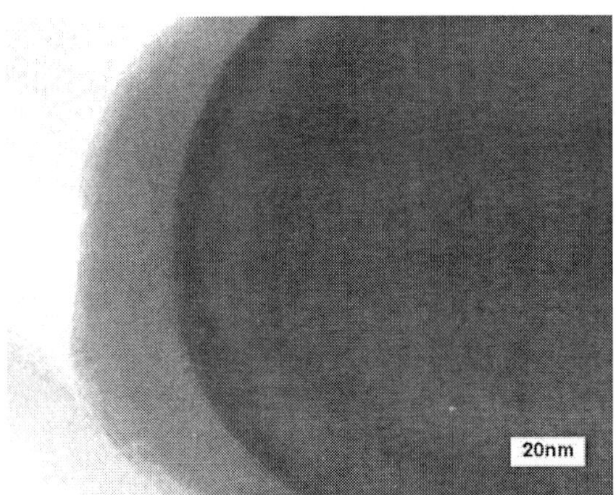

Figure 8. TEM picture of TiO_2 primary particle aqueous dispersion stabilized by EHEC (15 wt.-% relative pigment) after ultrasonic treatment.

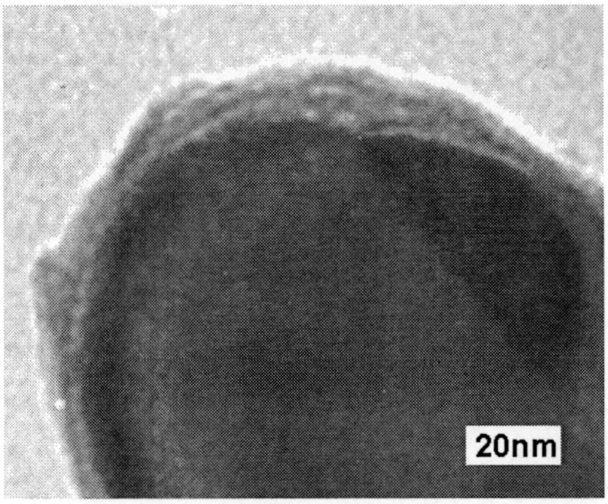

Figure 9. TEM picture of Fe_2O_3 primary particle aqueous dispersion stabilized by EHEC (1 wt.-% relative pigment) without mechanical (ultrasonic) treatment.

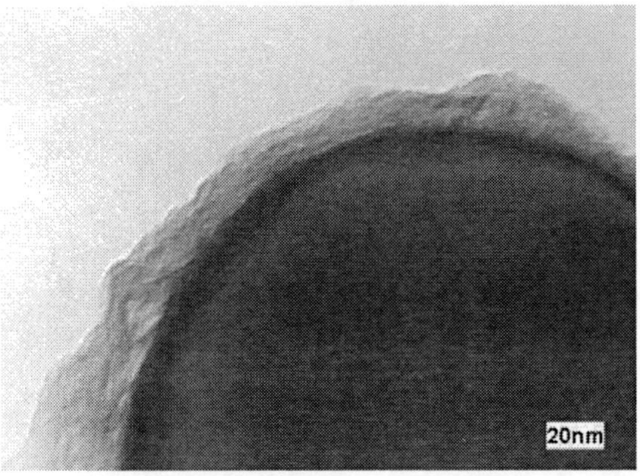

Figure 10. Picture of Fe_2O_3 aqueous dispersion stabilized by EHEC (7 wt.-% relative to pigment) after ultrasonic treatment.

Thus, the TEM investigations allow to conclude that mechanical, i.e., ultrasonic treatment has a pronounced effect on the pigment-polymer interaction in the dispersed system: the amount of polymer adsorbed and the thickness of the adsorption layer is significantly increased due to the activation of the particle surface.

In order to further elucidate the origin of this polymer adsorption layer growth, ESA measurements under variation of the frequency of the applied alternating electrical field have been carried out. These data can be used to get quantitative information about the structure of the adsorption layers according to the polymer gel layer theory (cf. [15]). In this theory it is assumed that the polymer adsorption layer consists of an inner layer of polymers adsorbed on the particle surface and an outer layer formed by the interpolymer interactions; the inner layer has a higher density and its formation is a direct result of pigment-polymer interactions.

The procedure given in literature [16] for the calculation of the thicknesses of the inner (Δ) and outer (δ) polymer adsorption layers after the polymer gel layer theory was applied for non-treated and ultrasonically treated dispersions of TiO_2 and Fe_2O_3, stabilized by EHEC. The formula for calculating the relative dynamic mobility μ/μ_0 (ratio of the dynamic mobility μmeasured for polymer containing pigment dispersion to the dynamic mobility μ_0 of the polymer free dispersion) is given in Eq. (8):

$$\frac{\mu}{\mu_0} = \ell^{-\kappa\Delta}\left\{-\frac{i\omega\alpha/\gamma-k^2}{k^2-\kappa^2} * \left[\ell^{-\kappa\delta}\left(1+\frac{\kappa}{k}\tanh k\delta\right)-\frac{\kappa^2}{k^2\cosh k\delta}\right]-\frac{i\omega\alpha}{\gamma k^2}\right\} \tag{8}$$

Here the quantity k is given by:

$$k^2 = \frac{\alpha}{\eta}\left(1+\frac{i\omega\eta}{\gamma}\right) \tag{9}$$

and the quantity γ by:

$$\gamma = \omega_0\eta \tag{10}$$

with Δ – thickness of inner adsorption layer, δ – thickness of outer adsorption layer, ω – frequency, η – dynamic viscosity, κ – Debye-Huckel parameter, α – drag coefficient, ω_0 – relaxation frequency.

In our case one has to take into account the differences in the particle size and particle inertia factor in the dispersion which was proven not to be the same for both the ultrasonically treated and non-treated samples without and with the addition of EHEC polymer. For ultrasonically not-treated samples, the difference is due to the disjoining forces (Deryagin's effect), for treated samples this deflocculation results from both the ultrasonic treatment as well as the stronger polymer adsorption which decreases of particle interaction strength (Rebinder's effect). The inertia factor G(r) can be calculated by the formula 11:

$$G(r) = \frac{1+(1+i)\sqrt{r/2}}{1+(1+i)\sqrt{r/2}+i(r/9)(3+2(\Delta\rho/\rho_m))} \tag{11}$$

where r is particle radius, ρ_m is the density of the solvent, Δρ is the difference between densities of the particles and the solvent

First, the relative dynamic mobility, i.e., the ratios between experimentally obtained values of the dynamic mobility of pigment particles covered by polymer and those of non-covered particles were established for the whole frequency range (data points in Figure 11 – 14). Then, using the equation (8), the thicknesses of inner (Δ) and outer (δ) adsorption layers were calculated. These data are compiled in Table 4. Inserting averaged values into the formula (8) one gets the theoretical dependency of the relative dynamic mobility (ratio between dynamic mobilities of polymer-coated and pure particles) from the frequency of the applied electrical field (lines in Figure 11 – 14). The comparison of the experimental and theoretical data allows to check the correctness of the calculations; it is evident that for all cases the theoretical curves were in good agreement with the experimental data.

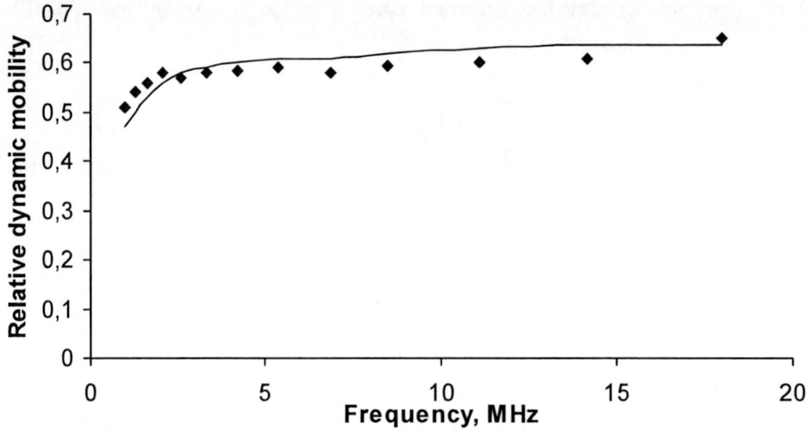

Figure 11. Dependence of the relative dynamic mobility on frequency for aqueous TiO_2 dispersions stabilized by EHEC (1 wt.-% relative to pigment) without ultrasonic treatment. ◆: experimental data, solid line: calculated theoretical curve.

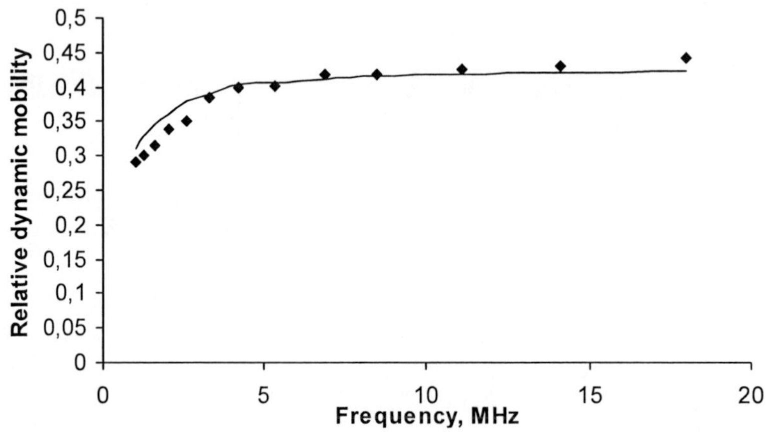

Figure 12. Dependence of the relative dynamic mobility on frequency for aqueous TiO_2 dispersions stabilized by EHEC (15 wt.-% relative to pigment) without ultrasonic treatment. ◆: experimental data, solid line: calculated theoretical curve.

The total thickness of the EHEC polymer adsorption layer as given by the sum of the inner and outer adsorption layer as calculated by applying the gel layer theory (see data Table 4) is in fair agreement with the values based on the polymer saturation concentration established by the concentration dependency of the ESA signals (see Table 3). The comparison the values of thicknesses of the inner and outer adsorption layers of the dispersion without and with ultrasonic treatment reveals that for both systems investigated the ultrasonic treatment predominantly effects the thickness of the inner layer which is especially prominent for the TiO_2 pigment. This reflects the activation of the pigment surface by the ultrasonic power leading to an increased adsorption of polymer; on the other hand, the outer layer which results from interpolymer interactions is less effected and maintains relatively unchanged in thickness.

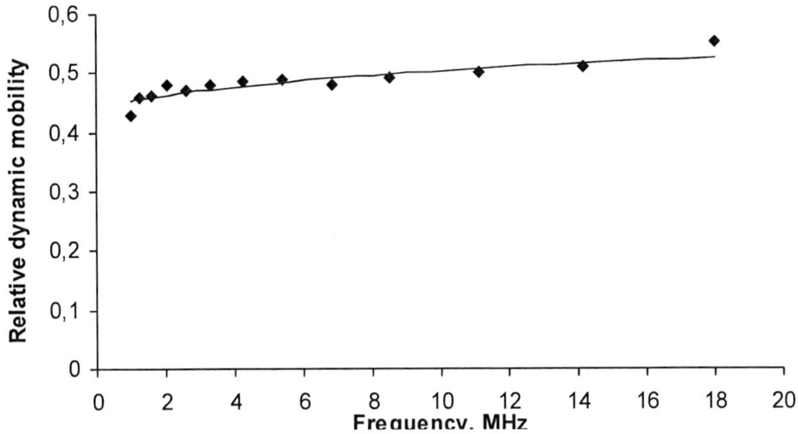

Figure 13. Dependence of the relative dynamic mobility on frequency for aqueous Fe_2O_3 dispersions stabilized by EHEC (1 wt.-% relative to pigment) without ultrasonic treatment.◆: experimental data, solid line: calculated theoretical curve.

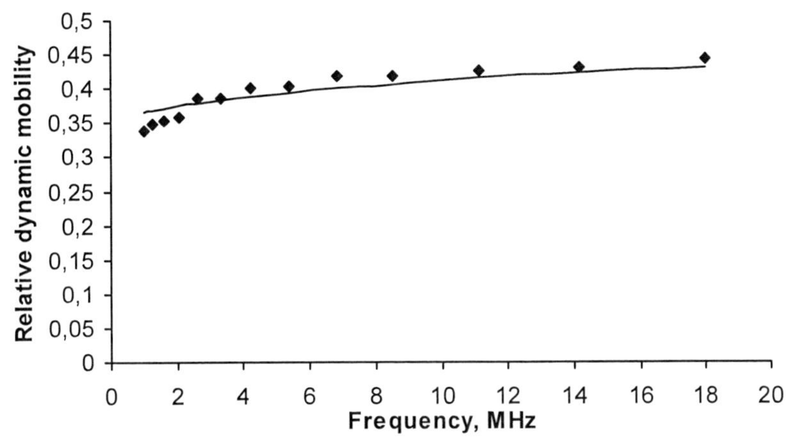

Figure 14. Dependence of the relative dynamic mobility on frequency for aqueous Fe_2O_3 dispersions stabilized by EHEC (7 wt.-% relative to pigment) without ultrasonic treatment.◆: experimental data, solid line: calculated theoretical curve.

Table 1. Effect of ultrasonic treatment on particle diameter (d_{part}) and increase of particle surface area(S_{tot}) in comparison to the saturation concentration (SC) of polymer adsorption; indices 1, 2: non-treated and ultrasonically treated sample

System	Averaged particle diameter, μm		Ratio between particle surface areawith and without ultrasonic treatment $S_{tot\,2}/S_{tot\,1}$	Ratio between saturation concentration with and without ultrasonic treatment SC_2/SC_1
	Without ultrasonic treatment	After ultrasonic treatment		
TiO_2 + EHEC	0.85	0.25	3.4	15
Fe_2O_3 + EHEC	1.4	0.3	4.7	7

Table 2. Ratios between saturation concentration (SC) and corresponding change of dynamic mobilityΔμ for pigment dispersion without (index 1) and with (index 2) ultrasonic treatment

System	Ratio between saturation concentration after and without ultrasonic treatment SC_2/SC_1	Difference between dynamic mobility of pure pigment and that at the saturation concentration without ultrasonic treatment $\Delta\mu_1$	Difference between dynamic mobility of pure pigment and that at the saturation concentration after ultrasonic treatment $\Delta\mu_2$	$\Delta\mu_2/\Delta\mu_1$
TiO_2 + EHEC	15	0.5	1.7	3.4
Fe_2O_3 + EHEC	7	1.7	2.5	1.47

Table 3. Thicknesses (d; cf. Eq. (6)) of the EHEC polymeradsorption layer on the TiO_2 and$Fe_2O_3$for treated and non-treated dispersions as calculated on the basis of the saturation concentration obtained from the ESA measurements

System	Thickness of adsorption layer without ultrasonic treatment, nm.	Thickness of adsorption layer after ultrasonic treatment, nm.	Ratio between thicknesses of treated and non-treated samples
TiO_2 + EHEC	5	17	3.4
Fe_2O_3 + EHEC	11	16	1.5

Table 4. Thickness of inner (Δ) and outer (δ) EHEC polymeradsorption layer on TiO_2 andFe_2O_3pigment for dispersions without and with ultrasonic treatment as calculated from Eq. (8) by using experimental ESA data (see text)

System	Thickness (Δ) of the inner adsorption layer in nm		Thickness (δ) of the outer adsorption layer in nm	
	without ultrasonic treatment	with ultrasonic treatment	without ultrasonic treatment	with ultrasonic treatment
TiO_2 + EHEC	1.2	8.7	4.7	6.5
Fe_2O_3 + EHEC	2.4	3.8	7.3	7.7

CONCLUSIONS

ESA and TEM measurements have shown that the improved colloidal stability of ultrasonically treated dispersed pigment/polymer stabilizer systems as known from sedimentation studies is due to special ultrasonic treatment related pigment-polymer interactions. For aqueous TiO_2 and Fe_2O_3 dispersions stabilized by EHEC, the saturation concentrations of polymer surfactants, i.e., the maximum amount of adsorbed polymer were established, and the increasing of the thickness of the adsorption layers upon ultrasonic treatment was proven by ESA measurements and confirmed by TEM investigations of the particulate systems.

By means of the ESA technique insights into the ultrasonically induced change of the thickness and structure of the polymer adsorption layers were obtained by applying the gel layer theory: In particular, the data allow to conclude that ultrasonic treatment of aqueous inorganic pigment dispersions is a powerful method for pigment surface modification leading first to an activation of the pigment surface; as a consequence, improvement of the pigment-polymer interaction is achieved which results in the creation of polymer adsorption layers of high thickness.

Thus ultrasonic treatment enables to obtain stable and high-dispersed pigment suspensions with modified pigment surface; this method opens new perspectives to modify the pigment surface more efficiently.

REFERENCES

[1] "*Surfactants and Polymers in Aqueous Solution", 2nd* edition, K. Holmberg, B. Jönsson, B. Kronberg, B Lindman, J. Wiley and Sons, Chichester, UK 2003.

[2] R.R. Netz, D. Andelman. Polymer adsorption in aqueous media, *Physics Reports,* 380 (2003) 1-95.

[3] P. Somasundaran, S. Krishnakumar. Adsorption of surfactants and polymers at the solid-liquid interface, Colloids and Surfaces, A: *Physicochemical and Engineering Aspects,* 123(1997)491-513.

[4] Theo G.M. Van de Ven. Surface modification of pigments with temperature-responsive polymer grafted by plasma-induced polymerization,*Advances in Colloid and Interface Science,* 48 (1994) 121-140.

[5] P. Somasundaran, Xiang Yu. Surface modification strategies for multicomponent polymer systems, *Advances in Colloid and Interface Science,* 53 (1994) 33-49.

[6] V. P. Zubov, I. F. Kuzkina, I. I. Ivankova, O. J. Schmitz, Effect of Ultrasonic Treatment on Stability of TiO_2 Aqueous Dispersions, *European Coating Journal,* 1998, 12, 954.

[7] V. P. Zubov, I. F. Kuzkina, I. I. Ivankova, O. J. Schmitz, Aqueous Dispersions of Copper Phthalocyanine, *European Coating Journal,* 1998, 11, 856.

[8] M. Antonietti, M. C. Weissenberger, Influences of monomer and oligomer structures in their adsorption on metal oxide surfaces,*Macromol. Rapid. Com.* 1997, 18, 295.

[9] V. P. Zubov, N. V. Serebryakova, I. A. Arutyunov, I. F. Kuzkina, N. A. Bulychev and Yu. A. Khrustalev, Effect of mechanical treatment of inorganic pigments surfaces on stability of their aqueous dispersions in the presence of ethylhydroxyethylcellulose, *Colloid Journal (Translated from Kolloidny Zhurnal)*, 2004, 66, 302.

[10] N. A. Bulychev, I. A. Arutunov, V. P. Zubov, B. Verdonck, T. Zhang, E. J. Goethals, F. E. Du Prez, Block Copolymers of Vinyl Ethers as Thermo-Responsive Colloidal

Stabilizers of Organic Pigments in Aqueous Media,*Macromolecular Chemistry and Physics,* 2004, 205 (18), 2457-2463.

[11] N. A. Bulychev, V.N. Fomin, I. A. Arutunov, V. P. Zubov, O.V. Abramov, E.V. Kisterev, Effect of mechanical action on surface modification in aqueous pigment dispersions, *Materialovedenie*, 2005, №4, 20.

[12] C. D. Eisenbach, Ch. Schaller, T. Schauer, K. Dirnberger, Surface modification of pigments by adsorption of amphipolar copolyelectrolytes, ACS Symposium Series 881"Particle Sizing and Characterization", American Chemical Society, 2004, 215.

[13] C. Schaller, A. Schoger, K. Dirnberger, T. Schauer, C. D. Eisenbach, Stabilizing pigments in full-mixing-systems,*Macromol. Symp*. 2002, 179, 173.

[14] C. Schaller, T. Schauer, K. Dirnberger, C. D. Eisenbach, Synthesis and stabilizing properties of amphipolar polyelectrolytes, *Eur. Phys. J.* 2001, E 6, 365.

[15] R.W. O'Brien, Determination of Particle Size and Zeta-Potential,*J. Fluid Mech.*, 1988, 190, 71.

[16] M.L. Carasso, W.N. Rowlands, R.W. O'Brien, The Effect of Neutral Polymer and Nonionic Surfactant Adsorption on the Electroacoustic Signals of Colloidal Silica,*J. Colloid and Interface Science,* 1997, 193, 200-214.

In: Chemical and Biochemical Kinetics: New Perspectives ISBN 978-1-61209-224-9
Editor: Gennady Zaikov

Chapter 8

PIGMENT SURFACE MODIFICATION IN PIGMENT DISPERSIONS WITH THERMO-RESPONSIVE POLY(METHYLVINYLETHER) COPOLYMERS IN COMBINATION WITH ULTRASONIC TREATMENT

N. Bulychev,[1,2] E. Kisterev,[1] O. Confortini,[3] F. E. Du Prez,[3] V. Zubov,[4] and C. D. Eisenbach[2,5]

[1]N.S. Kurnakov Institute of General and Inorganic Chemistry of Russian Academy of Sciences, Moscow, Russia
[2]Institute for Polymer Chemistry, University of Stuttgart, Stuttgart, Germany
[3]Department of Organic Chemistry, Polymer Chemistry Research Group, Ghent University, Ghent, Belgium
[4]Lomonosov Moscow State Academy of Fine Chemical Technology, Moscow, Russia
[5]Research Institute for Pigments and Coatings, Stuttgart, Germany

ABSTRACT

The process of surface modification of hydrophobic organic pigments (copper phthalocyanine (CuPc) and carbon black) as well as a hydrophilic inorganic pigment (titanium dioxide) in aqueous dispersions by employing tailor-made thermo-responsive copolymers, and the colloidal stability have been studied. The pigment surface modification is achieved by conventional adsorption and by (thermo)precipitation of amphiphilic methyl vinyl ether (MVE) containing polyvinylether block and PMVE graft copolymers with poly(ethylene oxide) side chains exhibiting a lower critical solution temperature (LCST). The effect of mechanical treatment of the pigment dispersion by ultrasonic power alone or in combination with the LCST property was investigated. The course of the pigment surface coating process was followed by the Electrokinetic Sonic Amplitude (ESA) method. The temperature-controlled sorption of PMVE-g-PEO graft copolymers on both inorganic and organic pigment surface was investigated. It was found that ultrasonic treatment together with LCST thermoprecipitation is a promising method for the surface modification of pigments with regard to dispersion stability.

Keywords: thermo-responsive copolymers, ultrasonic treatment, aqueous dispersions, inorganic pigments, organic pigments, lower critical solution temperature

AIM AND BACKGROUND

In numerous investigations, a variety of block copolymers based on poly(methyl vinyl ether) (PMVE) has been already used for the stabilization of CuPc and carbon black aqueous dispersions. PMVE belongs to the class of thermo-responsive polymers and exhibits a lower critical solution temperature (LCST) at 37°C. The introduction of such a thermo-responsive segment in an otherwise temperature-insensitive polyvinylether provided an efficient strategy for the design of a new class of "intelligent" polymer surfactants with stabilization activity which is responsive to environmental conditions. The optimal surfactants showed stability properties that are even better than those of the commercial Pluronics. In this paper, the process of the adsorption of P(MVE-b-IBVE) block copolymers and poly(ethylene oxide) (PEO) side chain containing P(MVE-g-EO) graft copolymers on the surface of hydrophobic copper phthalocyanine (CuPc) or carbon black and hydrophilic TiO_2 particles in aqueous dispersion will be studied by advanced methods at various polymer concentrations.

INTRODUCTION

In recent years, polymer surfactants have been of increasing interest as stabilizers for particulate systems, especially for emulsions and suspensions. Aqueous colloidal dispersions of pigments are important, ecologically friendly colloidal systems widely used in polygraphic and paint industries. The pigment particles are usually of 20 – 200 nm diameter and may be quite hydrophobic. In order to achieve a good stabilization in aqueous pigment dispersions, many formulations were proposed [1 - 8]. Earlier we reported about the role of mechanical, e.g. ultrasonic treatment to obtain highly stable dispersions [7, 9]. It was shown that using of polymer surfactants in combination with ultrasonic action can significantly improve the quality of dispersed systems. However, some aspects concerning pigment-polymer interaction and formation of adsorption layers under mechanical action needed additional elucidation.

The colloid stabilization of aqueous dispersions with polymer surfactants is believed to be a consequence of adsorption of the amphiphilic macromolecules on the particle surface resulting in mono- or multi-layers of certain structure and thickness which provide certain sterical and/or electrostatic stabilization effects. Polymer adsorption from aqueous solution on a particle surface is a result of specific interactions of various active sites on the particle surface with corresponding sites (groups) of the macromolecule. Therefore the chemical structures of the stabilizers are believed to be adjusted to the nature (e.g. hydrophilicity, charge etc.) of each type of the particles.

Thermo-responsive "smart" polymers exhibiting lower critical solution temperature (LCST) in aqueous solutions change the hydrophilic-hydrophobic balance when passing through LCST (they become more hydrophobic above LCST). One can expect that this property may result in temperature-controlled sorption and structure of the adsorption layer, and thus allowing colloid stabilization of aqueous dispersions. In other words, temperature-controlled changes in the hydrophobic-hydrophilic balance of those "smart" polymers can be used to adjust surfactant behavior of the polymer in order to achieve colloidal stabilization effects for various particles without changes of chemical structure or composition of the stabilizer.

Detailed information about the interaction of copolymers with the pigment surface, e.g., adsorption/desorption phenomena as reflected from the dynamic mobility μ_D of the dispersed particle, can be obtained by applying the Electrokinetic Sonic Amplitude (ESA) technique [10 - 17]. With this method, a potential is measured from pressure waves generated by the movement of charged particles in an oscillating electric field. The variation of this potential with frequency can be used for determining μ_D of particles reflecting the formation of thin polymer layers on their surface and for calculating the zeta potential as well as the particle size. For fixed low frequencies as applied in this work, μ_D is affected by the surface charge of the particle only [16, 18 - 20]. The study of the change of the dynamic mobility μ_D of aqueous TiO_2 dispersion upon addition of amphipolar polyelectrolytes has shown that ESA is a powerful method to reveal the course of polymer-pigment interactions and how this is related to the molecular architecture of the employed polyelectrolytes [19, 20].

Experimental Part

Materials

As pigments, β-copper phthalocyanine (β-CuPc) with primary particle size 0.1 μm and carbon black (channel type) FW 285 with primary particle size 11 nm were chosen. The PMVE-copolymers of various well-defined structures, molecular weights and narrow polydispersities, were synthesized by "living" cationic polymerization techniques as described elsewhere [22].

Techniques

Colloidal stabilization of the aqueous dispersions was monitored by sedimentation measurements of 1 % dispersions of CuPc and carbon black. Block copolymers used for stabilization of CuPc are insoluble in water. Therefore, for the preparation of the aqueous dispersions, the polymers were predissolved in ethyl acetate (5 %) and then introduced to the aqueous phase together with the pigment (ethyl acetate:water volume phase ratio was 1:10) [22]. After mixing with a laboratory stirrer (700 rpm for 10 min), the system was treated with ultrasound for 2 min with an ultrasonic generator UZDN-2 or Branson Sonifier B-12 with actual power of 1.5 W/cm^2. The pigment-polymer interaction was investigated by electrokinetic sonic amplitude (ESA) method. ESA measurements were carried out as described earlier [18]. Transmission electron microscopy (TEM) investigations were carried out according to the standard procedure. The samples of the pigment dispersions were diluted with distilled water by a factor of 100, then dropped on the TEM grid of 3mm diameter and dried on air for 1 day. The TEM pictures were taken by using a Jeol 2000 FX device under the voltage of 200 kV.

RESULTS AND DISCUSSION

Pigment Surface Modification with Graft Copolymers

Sedimentation Studies

First, P(MVE-g-EO) graft copolymers with a PMVE backbone and PEO side chains have been investigated. These polymers are hydrophilic below the LCST of PMVE and become amphiphilic above the LCST of PMVE. Hydrophilic pigments (e.g. TiO_2) are known to need hydrophilic surfactants for stabilization in aqueous dispersions, whereas hydrophobic pigments (CuPc and carbon black) need amphiphilic polymers to achieve stabilization of their aqueous dispersions. This opens a perspective to use these PMVE graft copolymers as universal smart surfactants for surface modification of both polar inorganic and non-polar organic pigments by taking advantage of the temperature-controlled sorption.

The stabilizing activity of the series of PMVE-g-PEO polymers in TiO_2 and CuPc aqueous dispersions were investigated at different temperatures. The results are given in Table 1. The half-time of the sedimentation, i.e. the period of time which is necessary for the border between colored and transparent zones in sedimentation measurements to reach 50% of the test-tube height was chosen as a measure of the stabilization activity.

Below the LCST of PMVE all copolymers are hydrophilic and thus, as one can see from Table 1, with exception of G1 act as efficient stabilizers of TiO_2 aqueous dispersions.

Regarding the relative difference of the hydrophilicity of the PEO grafts and the PMVE backbone it is assumed that the more hydrophilic PEO primarily interacts with the hydrophilic TiO_2 surface; i.e., the PEO grafts act as anchors. As far as the interaction of the graft copolymer with the hydrophobic CuPc is considered, it is possible to assume a preferred interaction with the less hydrophilic PMVE backbone.

However, after increasing the temperature above the LCST of the PMVE block, fast coagulation occurs and the size of precipitating particles becomes about 7 - 8 μm. This must be attributed to changes in the conformation of the PMVE backbone upon exceeding of the LCST (collapse of coil) and to interaction phenomena, i.e., the PMVE cannot act as a sterical stabilizing element anymore. As a consequence, the more hydrophilic PEO grafts which act as anchors at the TiO_2 surface are immobilized and buried. This means a relative hydrophobization of the polymer layer around the TiO_2 particle and particle coagulation causing sedimentation might result. On the other hand, a subsequent cooling below the LCST in combination with ultrasonic action allows to redisperse the system. The stability of the obtained dispersions is comparable to that of non-heated samples. This means that the agglomerates were broken by the action of ultrasonic power and that the PMVE backbone returned to the previous swollen state.

The poor colloidal stabilization with G1 polymer which contains only low molecular PEO segments is probably due to its weak adsorption on the TiO_2 particles and may indicate that adsorption is mainly due to interaction of high molecular PEO segments with TiO_2 surface. So one can conclude that PEO side chains of the copolymer are responsible for pigment-polymer interaction causing adsorption of PEO grafts on the pigment particles surface whereas the PMVE backbone allows for steric stabilization.

Table 1. Sedimentation stability of TiO_2 and CuPc aqueous dispersions with PMVE-g-PEO graft copolymers as a function of temperature

Code	Number of PEO grafts/Mn	Half-time of sedimentation			
		20 °C		50 °C	
		TiO_2, days	CuPc, min.	TiO_2, min.	CuPc, days
G1	9/2000	5	–	–	7
G2	3/5000	30	↑	↑	18
G3	3/5000	30	10 – 15	30 – 40	20
G4	7/5000	25	↓	↓	23
G5	13/5000	25	–	–	20

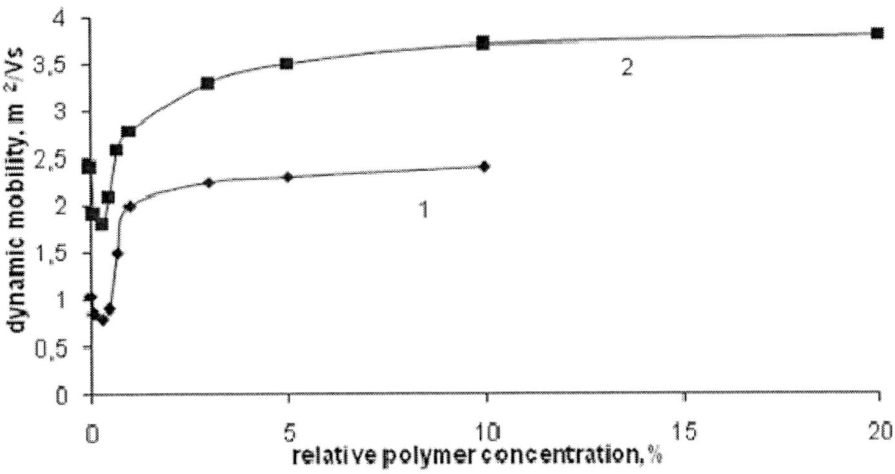

Figure 1. Dependence of the dynamic mobility on the relative concentration of PMVE-g-PEO (G3) for TiO_2 aqueous dispersion without and after ultrasonic treatment (temperature20 °C).

In the case of CuPc aqueous dispersions, reverse stabilization behavior is observed: above the LCST, all polymers act as stabilizers of CuPc aqueous dispersions, while fast coagulation is observed after temperature decrease to temperature below LCST.

This is attributed to the function of the collapsed PMVE backbone above the LCST as anchor to the CuPc surface. Comparing the hydrophilicity of PEO grafts and PMVE backbone below and above LCST, it is meaningful to assume that the collapsed PMVE above LCST is less hydrophilic than the expanded PMVE coil below LCST. This means that the exceeding of the LCST results in the formation of an amphiphilic copolymer with hydrophilic PEO coil and comparatively hydrophobic collapsed PMVE. As a consequence, above LCST the CuPc dispersion is stabilized by the interaction of "hydrophobic" collapsed PMVE with the CuPc surface.

When subsequently lowering the temperature below LCST, the adsorption of the graft copolymer on the CuPc surface is significantly weekend because of the dissolution (expanding) of the PMVE part of the graft copolymer which acts as anchor in the collapsed state only.

For both dispersion types, the best results are obtained for the graft copolymers with the longest PEO grafts (5000 vs. 2000). On the other hand, the number of grafts has not a significant influence on the sedimentation stability within the investigated range (between 3 and 13).

Electrokinetic Sonic Amplitude (ESA) Measurements

The adsorption of the PMVE-g-PEO copolymers on the TiO_2 and CuPc surface was investigated by the ESA method. For these measurements, the copolymer G3 (acting as the best stabilizer) was chosen (Figures 1 – 2). Other copolymers showed similar results.

From Figure 1 it is evident that the pigment-polymer suspensions show quite different behavior as a function of their preparation method. Without ultrasonic action, saturation of the surface adsorption layer seems to be reached at 1 % of PMVE-g-PEO in relation to the pigment TiO_2. After ultrasonic action, initial dynamic mobility of the TiO_2 is much higher, and saturation concentration of polymer is only reached at about 5 %. The differences in the initial values of the dynamic mobility and following differences in its changes due to the sorption of the polymer can be attributed to mechanical activation of the surface with ultrasonic treatment. Further addition of polymer did not effect the dynamic mobility. Thus, comparing the saturation concentrations for mechanically treated and non-treated samples, one can conclude that the ultrasonic action not only increases the pigment surface due to the obtained finer dispersion, but also leads to a higher thickness of the adsorption layer in comparison to the non-treated samples.

As already described above, fast coagulation of the TiO_2 dispersion was observed after exceeding the LCST of PMVE (Table 1). ESA examination of these coagulated dispersions enabled to prove that at the saturation concentrations, the dynamic mobility and ξ-potential of the particle surface did not change noticeably in comparison to those at room temperature; this indicates the absence of polymer desorption. This is also further prove of the earlier statement that pigment-polymer interaction at room temperature is apparently due to the adsorption of the PEO chains, while the PMVE backbones are preferably exposed into the aqueous media.

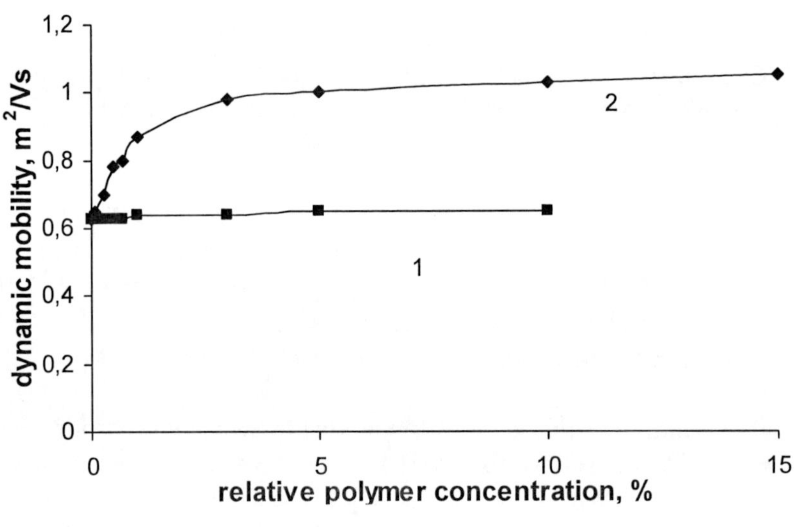

Figure 2. Dependence of the dynamic mobility on the relative concentration of PMVE-g-PEO (G3) for CuPc aqueous dispersion after ultrasonic treatment (temperature50°C).

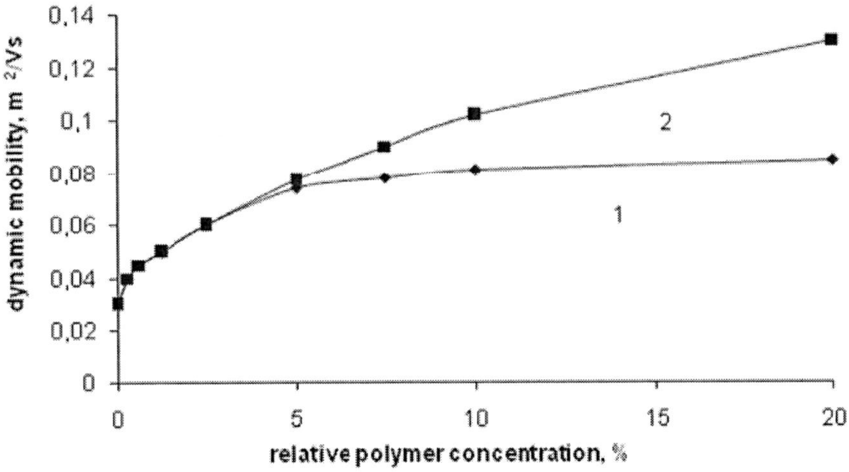

Figure 3. Dependence of dynamic mobility of surface treated CuPc on the relative concentration of $PMVE_{55} - PIBVE_{10}$ in aqueous dispersion in combination with ultrasonic treatment for pigment surface modification. 1 – conventional procedure, 2 – LCST thermoprecipitation.

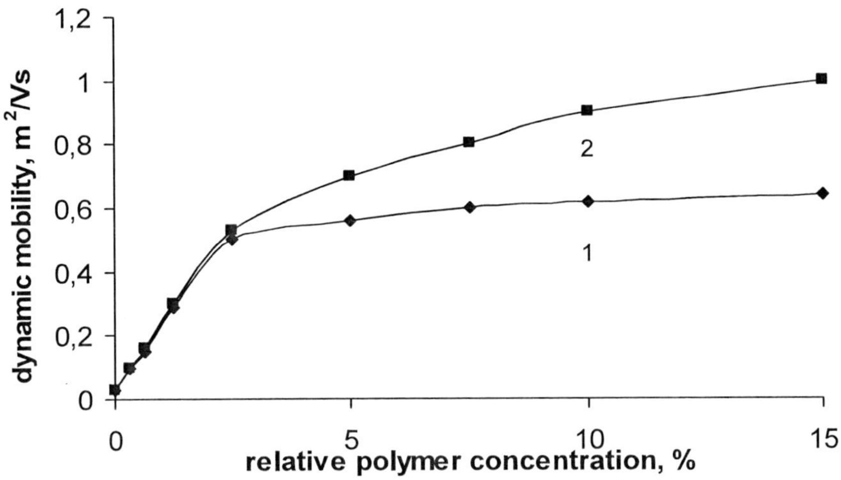

Figure 4. Dependence of dynamic mobility of surface treated CuPc on the relative concentration of $PIBVE_{22} - PMVE_{75} - PIBVE_{22}$ in aqueous dispersion in combination with ultrasonic treatment for pigment surface modification. 1 – conventional procedure, 2 – LCST thermoprecipitation.

Preliminary experiments have shown that hydrophilic polymers don't interact with hydrophobic CuPc surface in aqueous media. In case of PMVE-g-PEO copolymers, ESA measurements proved the absence of adsorption below LCST (Figure 2, curve 1). However as it is shown by the curve 2, above LCST of PMVE, the copolymer PMVE-g-PEO is able to adsorb on the CuPc surface in aqueous dispersions that results in high stabilization of CuPc dispersion. Saturation of the adsorption layer is reached at about 3 %. After decreasing of the temperature, the values of dynamic mobility and ξ-potential are similar to those for pure CuPc, which indicates the existence of desorption processes. This observation is in agreement with the fast coagulation process that was mentioned earlier.

Pigment Surface Modification of CuPc with Block Copolymers

ESA Studies

AB and ABA block copolymers of methylvinyl and isobutylvinyl ethers (PMVE–PIBVE and PIBVE–PMVE–PIBVE) exhibit a quite different thermo-responsive behavior as compared to the graft copolymers. They are amphiphilic below the LCST and become quite hydrophobic above the LCST. This gives a possibility to study their stabilizing properties in aqueous dispersions of organic pigments below and above LCST and to gain some insight into the pigment-polymer interaction and structure of the polymer layers in these systems at various temperatures.

First, aqueous dispersions of CuPc stabilized by $PMVE_{65}$ – $PIBVE_{10}$ and $PIBVE_{22}$ – $PMVE_{75}$ – $PIBVE_{22}$ (as the best colloidal stabilizers at room temperature [21]) were studied by ESA measurements (Figures 3 and 4).

Figures 3 and 4 represent the changing of the dynamic mobility of surface modified CuPc dispersion with increasing amount of polymer added to the system in the surface modification. As one can see, the pigment-polymer suspensions, stabilized by di- and triblock PMVE containing copolymers below the LCST (curves 1) show quite similar behavior. The saturation of surface layer seems to be reached at 5 % of $PMVE_{55}$ – $PIBVE_{10}$ and 3 % of $PIBVE_{22}$ – $PMVE_{75}$ – $PIBVE_{22}$ in relation to the pigment. Further addition of block copolymer did not noticeably increase the dynamic mobility. That means that formation of thick polymolecular layers hardly occurs. The carbon black dispersions showed similar behavior.

As PMVE in aqueous solutions exhibits a lower critical solution temperature (LCST = 37°C), it was of interest to study the creation of a surface layer by the LCST technique in combination with ultrasonic treatment, which may increase the thickness of the layers and stabilization of the dispersion in comparison to the polymer deposition below LCST. The LCST technique alone has been shown to be quite efficient in pigment surface modification [23].

In this work, the procedure of dispersion preparation was slightly changed. First, in constant temperature experiments (below LCST), the isothermal condition during the ultrasonic treatment is maintained by water cooling. In comparative experiments carried out without cooling, the heat generated by the ultrasonic treatment caused the temperature of the system to rise up to 45 – 50°C. As this temperature is above the LCST of PMVE, the block copolymers of PMVE became hydrophobic and were completely precipitated on the particle surface. The thickness of the deposited layers indeed increased (Figures 3 and 4, curves 2) and saturation concentration could not be established due to the thermoprecipitation (curves 2), which is ascribed to the fact that all polymer added to the solution precipitates on the particle surface. Thus, the more polymer is in the system, the thicker is the deposited layer, as is reflected by the higher values of dynamic mobility. This procedure gives the possibility to create surface polymer layers with controlled thickness.

Figure 5. Picture of CuPc particle in aqueous dispersion stabilized by $PMVE_{55}$ – $PIBVE_{10}$ after ultrasonic treatment. Scale 20 nm.

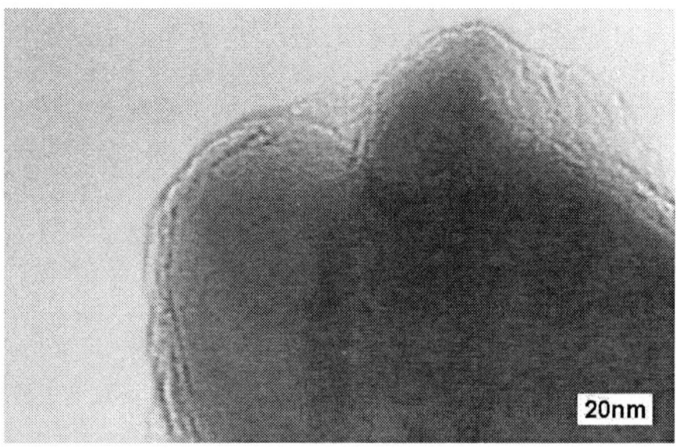

Figure 6. Picture of CuPc particle in aqueous dispersion stabilized by $PIBVE_{22}$ – $PMVE_{75}$ – $PIBVE_{22}$ after ultrasonic treatment. Scale 20 nm.

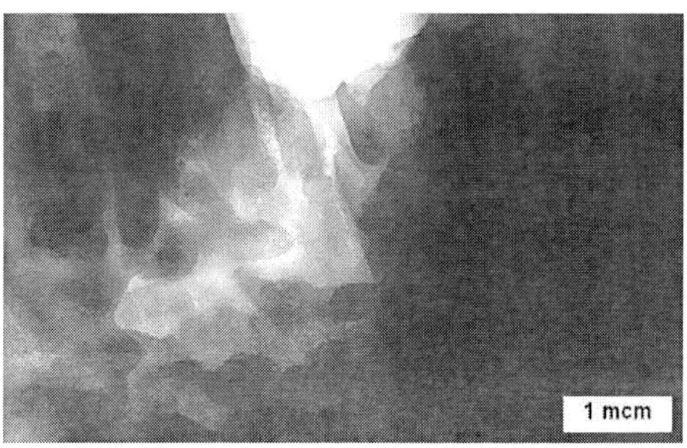

Figure 7. Picture of CuPc aqueous dispersion stabilized by $PMVE_{55}$ – $PIBVE_{10}$ after ultrasonic treatment according to LCST-thermoprecipitation technique. Scale 1 μm.

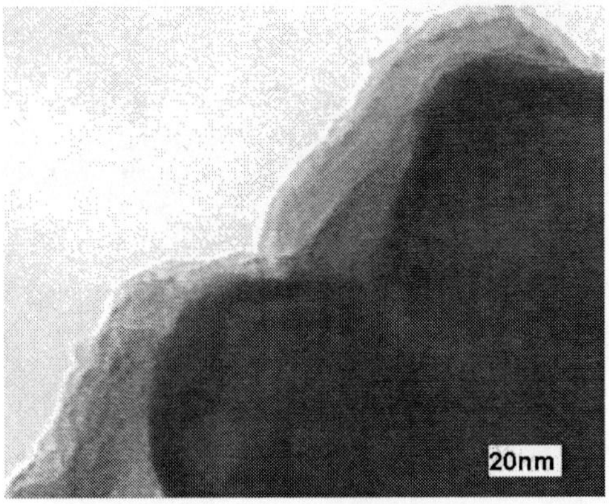

Figure 8. Picture of CuPc particle in aqueous dispersion stabilized by $PMVE_{55} - PIBVE_{10}$ after ultrasonic treatment according to LCST-thermoprecipitation technique. Scale 20 nm.

Transmission Electron Microscopy (TEM) Studies

In order to visualize the surface modification of the pigments in aqueous dispersion by treatment with ultrasound in the presence of LCST-polymers, the aqueous dispersions were investigated by transmission electron microscopy (TEM).

The pictures of CuPc particles stabilized by di- and triblock PMVE copolymers after mechanical (ultrasonic) treatment only are depicted in Figures 5 and 6. One can identify the polymer layers (grey zones) created by di- and triblock PMVE copolymers. The thickness of these layers is limited by the amount of polymer, which can be deposited on the particle surface up to saturation of polymer layer. The excess of the polymer will apparently create micelles and will not precipitate on the particles. As also seen from Figures 5 and 6, the thickness of the polymer layer created by triblock copolymer is higher as compared to the layer in the presence of the diblock copolymer. This is in agreement with the difference between the values of dynamic mobility obtained by ESA measurements (Figures 3 and 4).

In Figures 7 and 8, the TEM pictures of CuPc particles as covered by $PMVE_{55} - PIBVE_{10}$ according to the thermoprecipitation technique are given. From Figure 7 it is inferred that thermoprecipitation of the polymer, which becomes more hydrophobic above the LCST, leads to fast coagulation of the dispersion. The particles are strongly aggregated, with aggregate size of 5 – 10 μm. The polymer layer is much thicker in comparison to the layer thickness resulting from conventionally prepared systems (Figure 8). After decreasing of the temperature below LCST, the PMVE copolymer becomes amphiphilic and partial desorption of the polymer is observed; thus the system can be redispersed. In order to retain the thickness of the adsorption layer, one can consider a system in which the polymer can be crosslinked to prevent the desorption process (cf. [23]).

Conclusions

The temperature-controlled sorption of the PMVE-g-PEO copolymers on both inorganic and organic pigment surfaces was investigated. It was shown that depending on the temperature those polymers can adsorb on both hydrophilic and hydrophobic pigment surface. This gives a possibility to use these PMVE graft copolymers as universal smart surfactants for surface modification of both polar inorganic and non-polar organic pigments and for temperature-controlled interaction.

Block copolymers containing PMVE and hydrophobic blocks were shown to act as temperature-controlled stabilizers of CuPc, carbon black and TiO_2 aqueous dispersions. Mechanical treatment of dispersed pigment systems proven not only enhances the dispersion stability, but also influences the pigment-polymer interaction. It was further observed that the precipitation of polymers or polymer segments on the particle surface by exceeding the LCST is much more efficient than isothermal deposition below LCST for creation of thick polymolecular polymer layers on the particles surface.

REFERENCES

[1] "*Surfactants and Polymers in Aqueous Solution*", 2nd edition, K. Holmberg, B. Jönsson, B. Kronberg, B Lindman, J. Wiley and Sons, Chichester, UK 2003.

[2] R.R. Netz, D. Andelman. Polymer adsorption in aqueous media, *Physics Reports,* 380 (2003) 1-95.

[3] P. Somasundaran, S. Krishnakumar. Adsorption of surfactants and polymers at the solid-liquid interface, Colloids and Surfaces, A: *Physicochemical and Engineering Aspects,* 123(1997)491-513

[4] Theo G.M. Van de Ven. Surface modification of pigments with temperature-responsive polymer grafted by plasma-induced polymerization,*Advances in Colloid and Interface Science,* 48 (1994) 121-140

[5] P. Somasundaran, Xiang Yu. Surface modification strategies for multicomponent polymer systems, *Advances in Colloid and Interface Science,* 53 (1994) 33-49.

[6] V. P. Zubov, I. F. Kuzkina, I. I. Ivankova, O. J. Schmitz, Effect of Ultrasonic Treatment on Stability of TiO_2 Aqueous Dispersions, *European Coating Journal,* 1998, 12, 954.

[7] V. P. Zubov, I. F. Kuzkina, I. I. Ivankova, O. J. Schmitz, Aqueous Dispersions of Copper Phthalocyanine, *European Coating Journal,* 1998, 11, 856.

[8] M. Antonietti, M. C. Weissenberger, Influences of monomer and oligomer structures in their adsorption on metal oxide surfaces, *Macromol. Rapid. Com.* 1997, 18, 295.

[9] V. P. Zubov, N. V. Serebryakova, I. A. Arutyunov, I. F. Kuzkina, N. A. Bulychev and Yu. A. Khrustalev, Effect of mechanical treatment of inorganic pigments surfaces on stability of their aqueous dispersions in the presence of ethylhydroxyethylcellulose, *Colloid Journal (Translated from Kolloidny Zhurnal)*, 2004, 66, 302.

[10] R.W. O'Brien, Determination of Particle Size and Zeta-Potential, *J. Fluid Mech.*, 1988, 190, 71.

[11] R.W. O'Brien, B.R. Midmore, A. Lamb, R.J. Hunter, Interface characteristics in dispersed pigment systems. Faraday Discuss, *Faraday Discuss.Chem. Soc.,* 1990, 90, 301.

[12] M. Loewenberg, R.W. O'Brien, Electroacoustic Signals of Colloidal Silica, *J. Colloid Interface Sci.,* 1992, 150, 158.

[13] P.F. Rider, R.W. O'Brien, Dynamic Mobility of a Colloidal Particle, *J. Fluid. Mech.,* 1993, 257, 607.

[14] H. Maier, J.A. Baker, J.C. Berg, Elucidating pigment interactions with ESA, *J. Colloid Interface. Sci.,* 1987, 119, 512.

[15] N.P. Miller, J.C. Berg, Novel methods to determine particle sizes and interface characteristics in dispersed pigment systems, *Colloids and Surfaces*, 1991, 59, 119.

[16] M.L. Carasso, W.N. Rowlands, R.W. O'Brien, The Effect of Neutral Polymer and Nonionic Surfactant Adsorption on the Electroacoustic Signals of Colloidal Silica, *J. Colloid and Interface Science,* 1997, 193, 200-214.

[17] R.W. O'Brien, The Effect of a Adsorbed Polyelectrolyte Layer on the Dynamic Mobility of a Colloidal Particle, *Part. Syst. Charact.,* 2002, 19, 1.

[18] C. D. Eisenbach, Ch. Schaller, T. Schauer, K. Dirnberger, Surface modification of pigments by adsorption of amphipolar copolyelectrolytes, ACS Symposium Series 881"Particle Sizing and Characterization", American Chemical Society, 2004, 215.

[19] C. Schaller, A. Schoger, K. Dirnberger, T. Schauer, C. D. Eisenbach, Stabilizing pigments in full-mixing-systems, *Macromol. Symp*. 2002, 179, 173.

[20] C. Schaller, T. Schauer, K. Dirnberger, C. D. Eisenbach, Synthesis and stabilizing properties of amphipolar polyelectrolytes, *Eur. Phys. J.* 2001, E 6, 365.

[21] N. A. Bulychev, I. A. Arutunov, V. P. Zubov, B. Verdonck, E. J. Goethals, F. E. Du Prez, Block Copolymers of Vinyl Ethers as Thermo-Responsive Colloidal Stabilizers of Organic Pigments in Aqueous Media, *Macromol. Chem. Phys.,* 2004, 205 (18), 2457.

[22] E.J. Goethals, W. Reyntjens, X. Zhang, B. Verdonck, T. Loontjens, Thermoadjustable colloid stabilizers, *Macromol. Symp.* 2000, 157, 93.

[23] T. Schauer, C. D. Eisenbach, Organic polymer treatment - the way to modern pigments, *ECJ*, 2003, 3, 114.

In: Chemical and Biochemical Kinetics: New Perspectives
Editor: Gennady Zaikov ISBN 978-1-61209-224-9

Chapter 9

PROPERTIES OF CARBON-BLACK COMPOSITE VAPOR DETECTORS BASED ON MULTIFUNCTIONAL POLYMERS

Luciana Sartore,* Andrea Sassi and Marzia Barbaglio
Dipartimento di Ingegneria Meccanica e Industriale,
Università degli Studi di Brescia, Brescia, Italy

ABSTRACT

In this work the electrical properties of vapor detectors, formed from composites of conductive carbon-black and insulating organic multifunctional polymers having metal ions complexing ability, were investigated. The new composites are tailored to produce increased sensitivity toward specific classes of analyte vapors. Resonant frequency shift of a Quartz Crystal Microbalance (QCM) and dc resistance measurements have been also performed simultaneously on polymer-carbon black composite materials. For comparison purpose, poly(vinyl chloride) (PVC) with di(2-ethylhexyl)phthalate (DOP), a traditional low molecular weight plasticizer, is used as representative of the behavior of a traditional composite vapor detector.

These new detectors showed an enhanced sensitivity upon exposure to acetic acid and amines vapors; the performances of our systems are 10^3 times higher than those of a traditional composite vapor detector. Moreover the extent of such responses is beyond that expected by mass uptake upon exposure to the same vapors and cannot be attributed solely to differences in polymer/gas partition coefficients. In this respect, several different chemical factors determine the dc electrical response of this system: in our opinion changes in polymer conformation during the adsorption process also play a significant role.

The effects of the temperature on the electric resistance of the vapor detectors have been also studied. These materials showed a discontinuity in the temperature dependence of their resistance, and this discontinuity provided a simple method for determining the T_g of the composites.

* Corresponding author. E-mail address: luciana.sartore@ing.unibs.it.

Keywords: vapor detectors, multifunctional polymers, sensitivity, selectivity, composites, glass transition temperature

1. INTRODUCTION

Arrays of broadly responsive detectors have been used for many years as means of detecting a wide range of vapors. These systems have been proposed in several fields for the identification, classification and quantification of analyte vapors. Chemically sensitive resistors fabricated from conductive fillers dispersed into insulating organic polymers are an attractive implementation of this approach [1-4]. A variety of composites, actually, can be easily obtained and used to realize arrays of chemically different resistive detectors, using no expensive starting materials and equipment. The behaviour briefly suggests that they can be regarded as promising gas sensors characterized by easy processability and cost effectiveness.

The conductivity of these composite systems can be understood in terms of the percolation theory [5]. At low concentrations, generally, the filler incorporated in the form of small particles with a different shape, is distributed homogeneously in the volume of the insulating matrix, and there are no contacts between adjacent particles. With rising filler concentration, agglomerates of the filler particles start to form because the particles are in contact with each other. At a certain critical volume of this conductive filler, called percolation threshold, the growing agglomerates reach a size which makes possible to touch each other and in this way a continuous conducting network through the polymer matrix is formed. As a consequence of the first appearance of the network, the composite transforms from an insulator to a conductor.

Exposure of such a composite film to an organic vapors increases its resistivity because of the variation of the matrix volume fraction with respect to the conductive phase. Carbon black, Ag, Cu or conductive organic polymers have all been used as the conductive phase.

It is known from literature that another approach to obtain sensor diversity, involving modifications of the properties of a base polymer vapor detector, is attained through the addition of plasticizers. In such a way, several polymeric matrices and traditional low molecular weight plasticizers have been employed [6].

An interesting alternative might be represented by polymeric blends containing a low glass transition temperature (T_g) polymer, possessing high miscibility with the matrix [7]. The main advantage of such systems would be that drifting phenomena typical of low molecular weight plasticizers could be avoided. In addition, the properties of these sensors might be tuned by varying the plasticizer molecular weight.

The basic performance properties of such arrays, including detection limits and responses to mixtures of analytes have been described in literature [1, 2, 8-10].

Moreover recent studies are devoted to the development of chemiresistive vapor detectors with increased sensitivity towards analyte vapors [11-13].

In this context we studied new series of carbon black multifunctional polymers composites based on poly(amidoamine)s (PAAs) which show large enhancements in sensitivity towards specific classes of analytes such as amines and carboxylic acid vapors [14, 15]. In particular these materials have a sensitivity 10^3 higher than the performance of typical insulating polymer carbon black composite vapor detector. Compositional diversity in an array of such vapor detectors, can be achieved by varying the chemical structure of monomers. Nevertheless, having PAAs excellent metal ions complexing ability, diversity in the response behavior could also be obtained by chelating different heavy metal ions to the same carbon black polymer composite.

The carboxylic acid and amines vapor detection properties of PAA based materials, have been compared with the detection properties of lower polarity analytes, such as acetone and pentane.

For comparison purposes, a conducting carbon black-polymer composite detector based on poly(vinyl chloride) (PVC), as the base polymer, with the addition of di(2-ethylhexyl)phthalate (DOP), a typical low molecular weight plasticizer, was prepared and checked with all the analytes considered.

The goal of this work is to investigate whether it is possible, through the choice of the polymeric component of the carbon black polymer composite detector, to achieve enhanced sensitivity to a specific class of analyte vapors and to explore the mechanism by which these detectors achieve their enhanced sensitivity. Finally the effects of the temperature on the electric resistance of the vapor detectors have been also studied and it provided a simple method for determining the T_g of the composites.

2. Experimental

2.1. Materials

All the solvents used to generate analyte vapors were reagent grade and were used as received (Aldrich Co Milan, Italy); poly(vinyl chloride) (PVC) and di(2-ethilhexyl)phatate (DOP)were purchased from Aldrich Co (Milan, Italy).

2.2. Preparation of the Conductive Films

Two PAAs BAC-DMEDA and BAC-EDDA synthesized as previously described [14], and PVC plasticized with DOP were used in this work. The conductive films were obtained by spray from a suspension prepared as follows: 80 mg of organic matrix (PAAs or PVC with 40% by weight of DOP) were dissolved in 10 ml of solvent (water or tetrahydrofurane respectively) and 20 mg of carbon black were added. The mixture was sonicated for 5 minutes and therefore it was sprayed on a Quartz Crystal Microbalance (QCM) element until obtaining a frequency shift of about 20kHz. The films were dried in air for 24 hours before use.

The same procedure was followed to prepare films with PAAs complexing heavy metal ions. Those systems were obtained adding to the water solution, at pH 7.5, 15 wt% (by weight of the PAA) of Cu^{++} or Co^{++}.

Table 1. Polymer structure, viscosity, molecular weights and T_g

Structure	$\eta_{sp}/C^{a)}$(dl/g)	$\overline{M_w}$ (g/mol)	$\overline{M_n}$ (g/mol)	T_g (°C)
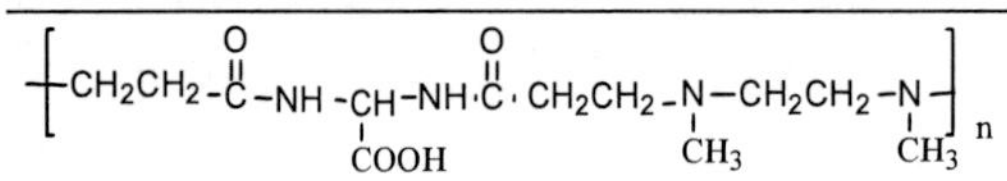 BAC-DMEDA	0.12	6050	4820	99
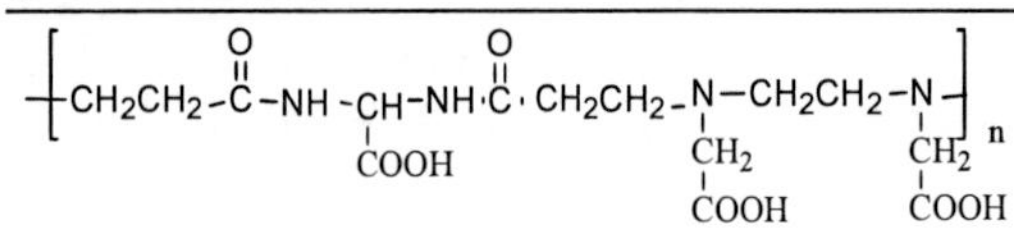 BAC-EDDA	0.10	6900	5900	122

a) C= 0.2% in NaCl 0.1 M.

2.3. Data Collection and Analysis

The calorimetric analysesof the compositeswere performed using a Digital Scanning Calorimeter (DSC) Q100 TA Instruments, interrelated with a TA Control Instrument Q Series.

QCM crystals (AT-cut, 6MHz, blank diameter = 13.5 mm, electrode diameter = 7.8 mm) with the custom electrode pattern were obtained from Elettroquarz S.r.l. (Milan, Italy). Frequency shift measurements were performed on carbon black polymer films in conjunction with simultaneous resistance measurements of these composite vapour detectors by means of an electrode configuration in which the standard QCM oscillation electrodes were at 90° angles to make room for two additional resistance tabs [8, 16]. The QCM technique is based on detection the mass at the surface of the quartz crystal by measuring the change of frequency ofthe crystal oscillation. The interaction of detectors with different analyte vapors on a surface of crystal resulted in increase of the mass of the crystal and in decrease of the oscillation frequency.

The measurements were carried out with a Laybold Inficon XTM/2 oscillator and frequency counter (East Syracuse, NY) and an APPA model 305 digital multimeter (Taiwan), both equipped with a LabView 6.0 acquisition system running on a Pentium III personal computer.

The frequency shift and the resistance response of each modified QCM sensor were measured as a function of time under exposition to different organic vapours. Each sensor was first placed into the sensing thermostatic chamber (volume = 0.604 L) equipped with a thermocouple, and then it was kept under a nitrogen stream (1 L/min) until steady state values of resistance and frequency were observed. Finally, the frequency and the resistance were sampled for 20 min more under static condition to obtain the baselines. All measurements were run in nitrogen, at 24°C, under static condition. All analytes were injected into the chamber and exposed to the detectors for 900 s, and then nitrogen was passed over the composite for 900 s maintaining a flow rate of 1 L /min. The adsorption and desorption cycles were repeated more times, increasing the vapour concentration in step of 4.9 ppth for acetone, 4.4 ppth for pentane, 6.5 ppm for 1-butylamine, 4.7 ppm for triethylamine and 1.14 ppm for acetic acid.

3. Results and Discussion

PAAs are regular amide polymers characterized by the presence of amido and tertiary amino groups regularly arranged along their macromolecular chain. They are obtained, in a linear form, with a considerable variety of structures by stepwise polyaddition of primary monoamines or bis-secondary amines, to bis-acrylamides [17]. PAAs are hydrophilic, water-soluble, generally characterized by moderately basic properties and high capacities to complex metal ions.

PAAs carrying additional functions (carboxyl, hydroxyl and amine) as side substituents can be easily obtained, starting from the appropriate monomers. These polymers have been extensively studied because the versatility of their chemical structures allows a variety of applications. In particular they were proposed as drug carrier in biomedical field, and in the environmental engineering for their heavy metal ions chelating properties [15, 17-19].

Two structurally related PAAs namely BAC-DMEDA and BAC-EDDA, which, in linear form, have been shown to possess the capacity to form stable complexes with well-defined stoichiometry, were selected for this study. The products were prepared as previously reported [14] and characterized in terms of molecular weights and molecular weight distributions by analytical Gel Permeation Chromatography (Table 1), making use of PAAs purposely prepared and analysed by NMR techniques.

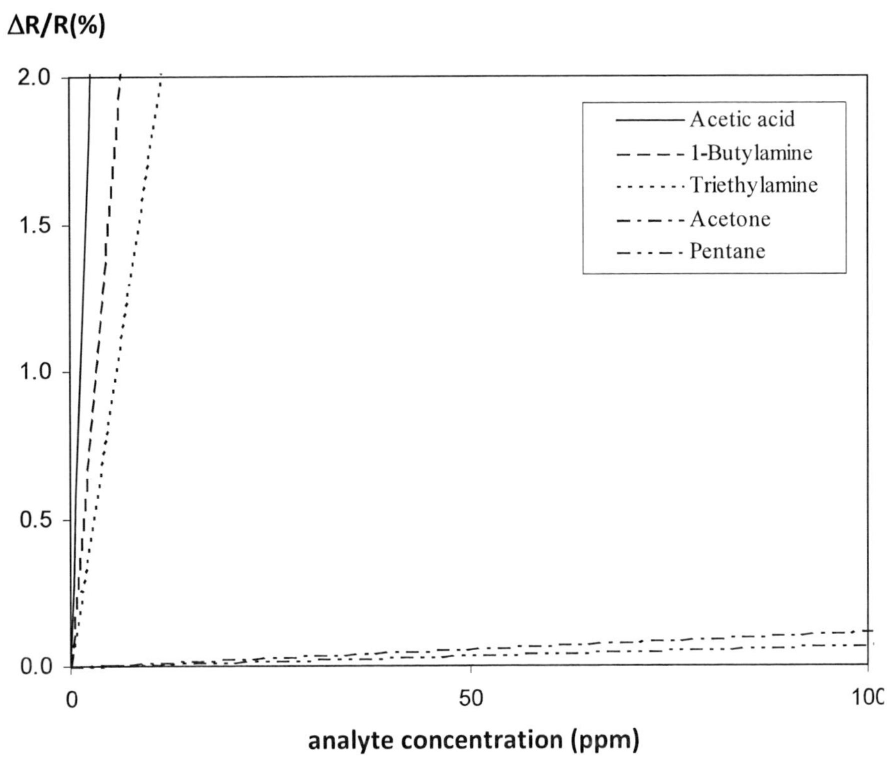

Figure 1. Normalized resistance increase upon exposure for 900 s (see experimental) of [BAC-DMEDA]Cu to increasing concentrations of different analytes.

The conductive composite detectors were obtained by spraying a suspension prepared by dissolving the polymer materials in water and adding the conductive filler. In order to explore the mechanism by which these detectors work, all sensors were tested upon exposure to different organic vapors using a modified QCM crystal in which the resonant frequency shift and dc resistance responses were performed simultaneously. Since the interaction between composite films with specific classes of analyte vapors on the surface of crystal produces an increase of the mass of the crystal,a corresponding decrease of the oscillation frequency was detected.

It was found that the new materials can detect the presence of gas by reversible variations in electrical properties. They showed high efficiency since they adsorb and desorb rapidly at room temperature and have a considerable endurance to continuous and repeated adsorption and desorption cycles.

Figure 1 presents the normalized resistance increase of a PAA-carbon black composite detector to various concentrations of volatile organic vapors having different polarity. The responses of new detectors showed high selectivity for different classes of analytes for example amines, alkanes or ketones. In particular the minimal amine as well as carboxylic acid vapour amount that they can detect is part per million (ppm), while the detection limit for other analytes with a lower polarity is part per thousands (ppth). This behaviour should be due to the chemical structure of the PAAs: both BAC-DMEDA and BAC-EDDA, actually, have a rather complex chemical structure with many polar groups arranged along the macromolecular chain and functional groups as side substituents which provide strong interactions with polar analyte vapors. Besides someanalytes which produced enhanced sensitivity are reported in Figures 2.A and 2.B, which show respectively the behaviour of BAC-EDDA and BAC-DMEDA. Each analyte produced a distinct response that was different from the response curve of every other analyte investigated. It was also possible to appreciate a diversification of the response curves against primary to tertiary amine. In this way it is possible to suppose a good selectivity of PAA-carbon black composite detectors even within the same analyte vapor family.

The behaviour of PAA composite film when exposed to above gases was compared to the relative differential resistance behaviour of PVC-DOP composite film, representative of the traditional materials used up to now for such application. The PAA composite detector clearly exhibited a significantly enhanced sensitivity as compared to other polymer composite detector films. Generally the new material can detect a mean value of amines or acetic acid concentration approximately 1000 fold smaller than that of PVC/DOP detector, in the same experimental conditions (prepared with the same filler and deposited in the same way on identical substrate) for the same analyte exposure.

As above reported, different chemical PAA structures also carrying additional functions as side substituents can be easily obtained starting from the appropriate monomers. These can be used as the insulating component to fabricate a variety of chemiresistive composite vapor detectors. Both PAA based detectors selected for this study showed enhanced sensitivity and selectivity to acetic acid and amines compared to other organic vapors.

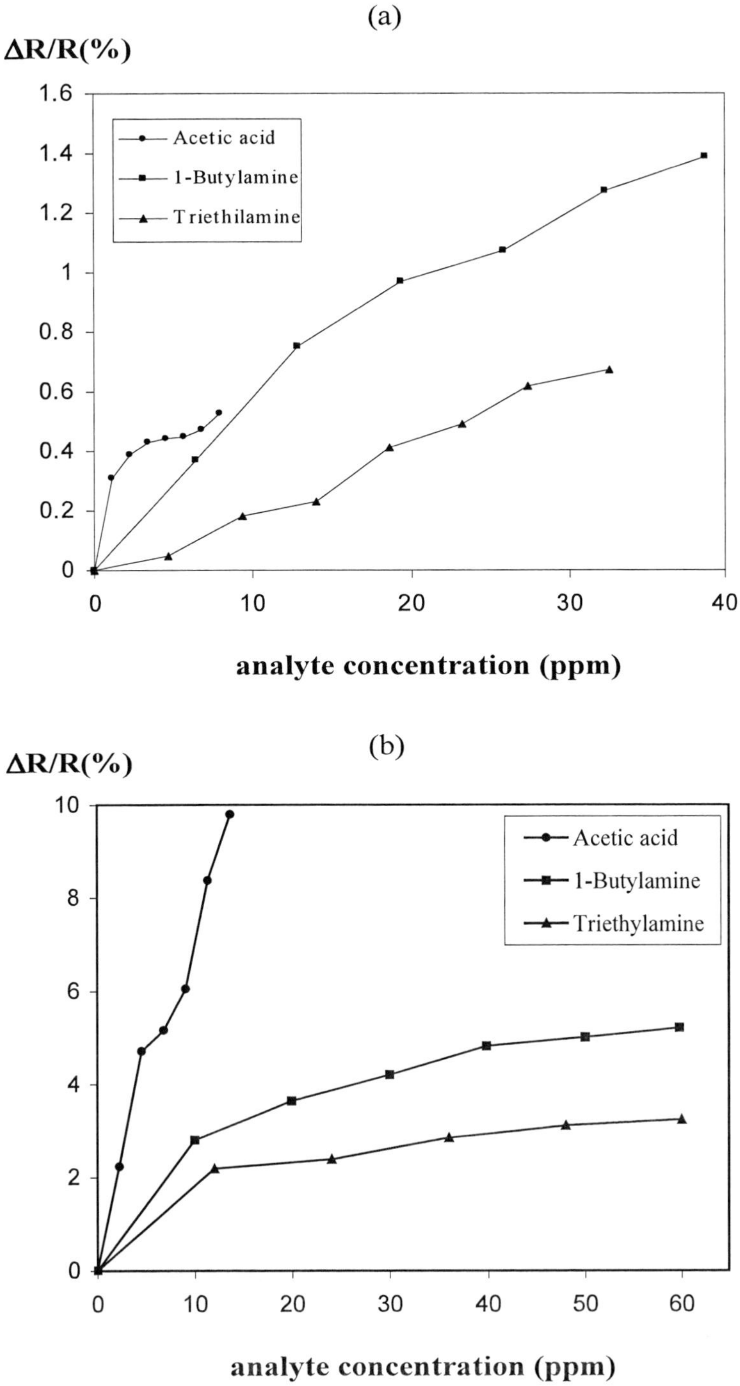

Figure 2. Comparison of the resistance responses to increasing concentrations of different analytes for: (a) [BAC-EDDA]Cu, (b) [BAC-DMEDA]Cu.

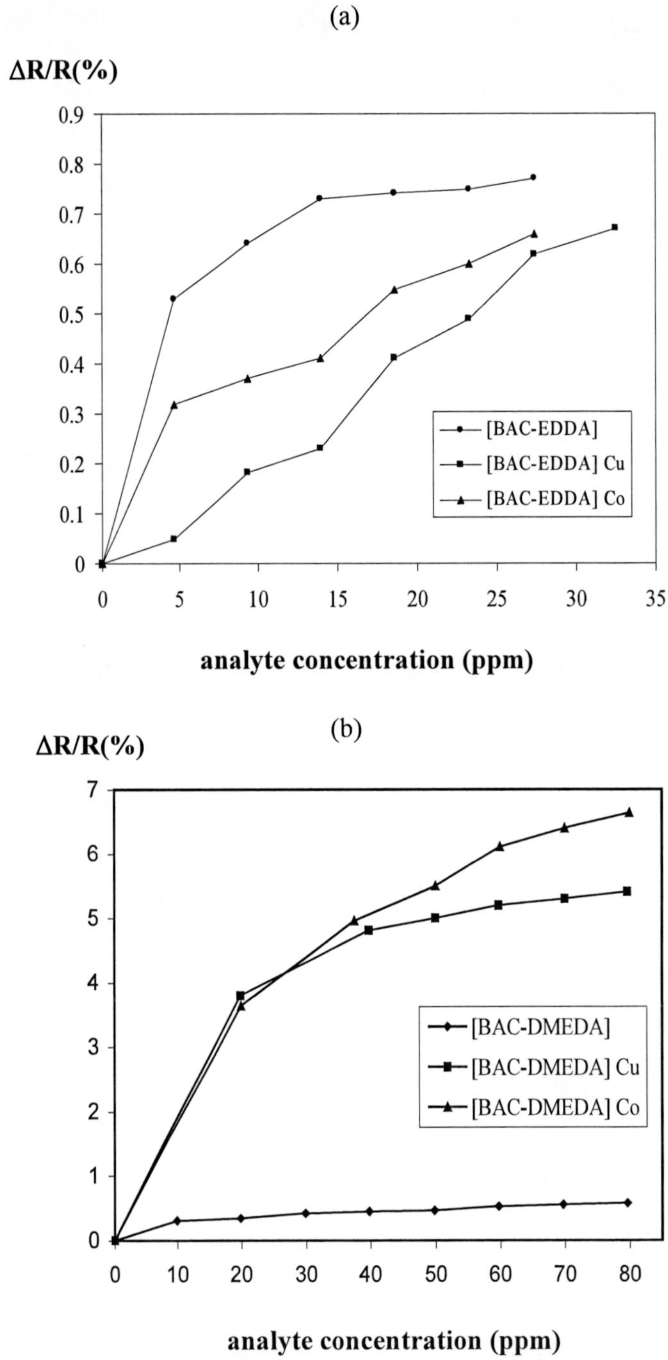

Figure 3. Comparison of the resistance responses under triethylamine vapour exposition for 900 s (see experimental) of: (a) [BAC-EDDA], [BAC-EDDA]Cu and [BAC-EDDA]Co; (b) [BAC-DMEDA], [BAC-DMEDA]Cu and [BAC-DMEDA]Co.

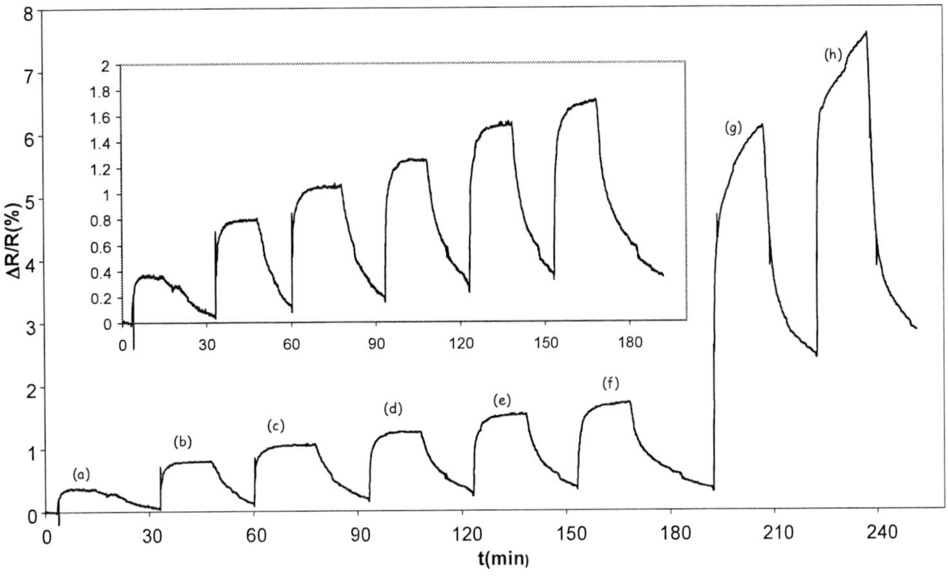

Figure 4. Resistance response upon sorption and desorption cycles carried out on [BAC-EDDA]Cu composite sensor using the following concentrations of 1-butylamine: (a) 6.5 ppm, (b) 13 ppm, (c) 19.5 ppm, (d) 26 ppm, (e) 32.5 ppm, (f) 39 ppm, (g) 1800 ppm, (h) 3600 ppm.

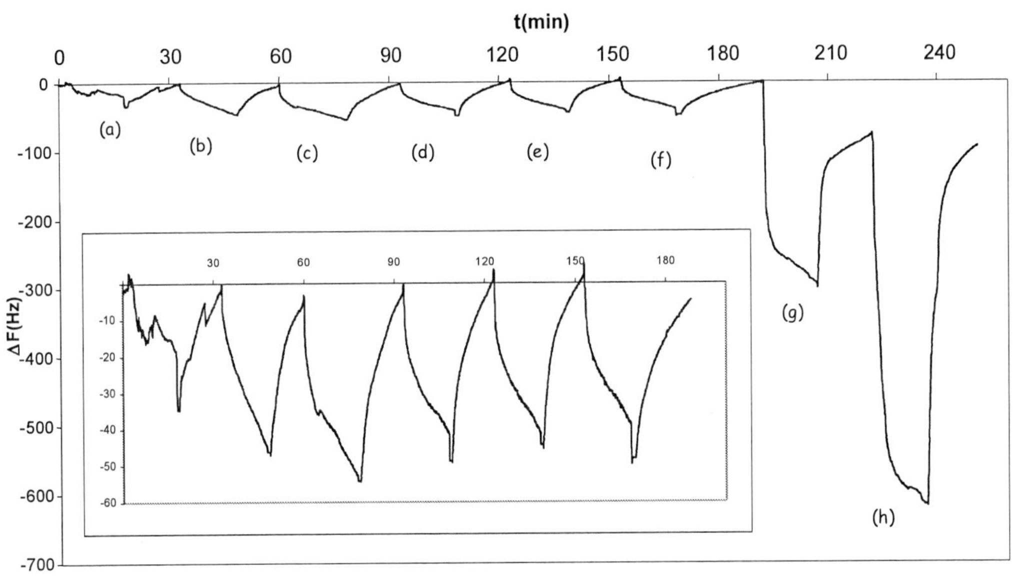

Figure 5. Frequency shift response under sorption and desorption cycles carried out on [BAC-EDDA]Cu composite sensor using the following concentrations of 1-butylamine: (a) 6.5 ppm, (b) 13 ppm, (c) 19.5 ppm, (d) 26 ppm, (e) 32.5 ppm, (f) 39 ppm, (g) 1800 ppm, (h) 3600 ppm.

Moreover a chemical diversification can be easily obtained by complexing different heavy metal ions on polymeric chains.In particular PAA's functional groups can act as ligands with heavy metal ions or provide strong interactions with other polar molecules. The complexation in water solution brings each monomeric unit to coordinate a metal ion, to give a more compact polymeric structure. The use of heavy metal ions produces chemical diversity in chemiresistive vapor detector improving the selectivity while maintaining the high overall sensitivity of the PAA based composites for specific classes of analytes. On this subject Figure 3 presents the normalized resistance increase upon exposure to triethylamine vapor for respectively [BAC-EDDA] (Figure 3.A) and [BAC-DMEDA] (Figure 3.B) polymer free and the same structures chelating Cu^{++} and Co^{++} as model ions. Each detector produced a distinct response that was differentiable from the response curve of the others investigated. Besides, figure 3.B shows that the presence of chelated metal ions on [BAC-DMEDA] allowed the sensors to amplify the response in terms of percentage variation of resistance.

The simultaneously measurements of QCM frequency shifts and resistance changes, were performed to determine the underlying cause for the enhanced sensitivity of PAA based materials to organic acids or amines compared to low polarity organic vapors. Figure 4 presents the resistance response of [BAC-EDDA]Cu vapor detector upon sorption and desorption cycles of 1-butylamine. The sorption steps were obtained by injection of increasing amount of analyte while nitrogen was passed over the detector to obtain desorption. The data indicate that the dc resistance change is directly related to the analyte concentration. Moreover these data fit well to a linear dependence of normalized resistance increase on analyte concentration also at ppm level as reported in insert of Figure 4 at an expanded scale.

The corresponding frequency shift measurements are reported in Figure 5. It can be observed that up to at least 40 ppm (cycles a through f) there is no significance trend in the frequency shift that can be related to the increase in analyte concentration. Only when the ppth level is reached (cycles g to h), the frequency shift shows a stepped response which scales with the concentration, demonstrating a detectable mass uptake in the cating film.

The above studies have demonstrated that the higher sensitivity to amines as well as acetic acid can not be attributed solely to the differences in polymer/gas partition coefficients, as it is for other organic vapors tested. The chemical structure of these polymers, characterized by the presence of specific groups regularly arranged along the macromolecular chain, strongly interact with target analytes. On the other hand this complex structure imparts rigidity and high organization in polymer packaging (i.e. hydrogen bonds). In this connection, we have recently demonstrated that the PAA ion chelation for example, occurring through a cyclic structure around the metal atoms, highly constrains polymer chains causing changes in molecular packing and suppressingsegmental motions and T_g[20]. Small amounts of analyte vapors probably disarrange these constraints causing changes in molecular packing and determine the enhanced sensitivity and selectivity observed.

Since temperature influences vapor sorption and permeability, that effect changes in the dc electrical resistance response of polymeric detectors [21], measurements of electrical resistance of PAA carbon black composites were performed also at several temperatures. All samples showed a pronounced change in resistance as the temperature was varied, in particular a decrease of the $\Delta R/R(\%)$ with the increasing temperature was observed. In addition a discontinuity in the temperature dependence of the resistance was also generally highlighted. For example Figure 6 displays the dc electrical resistance of [BAC-DMEDA]Co detector as a function of the temperature and the discontinuity formation approximately at 27°C.

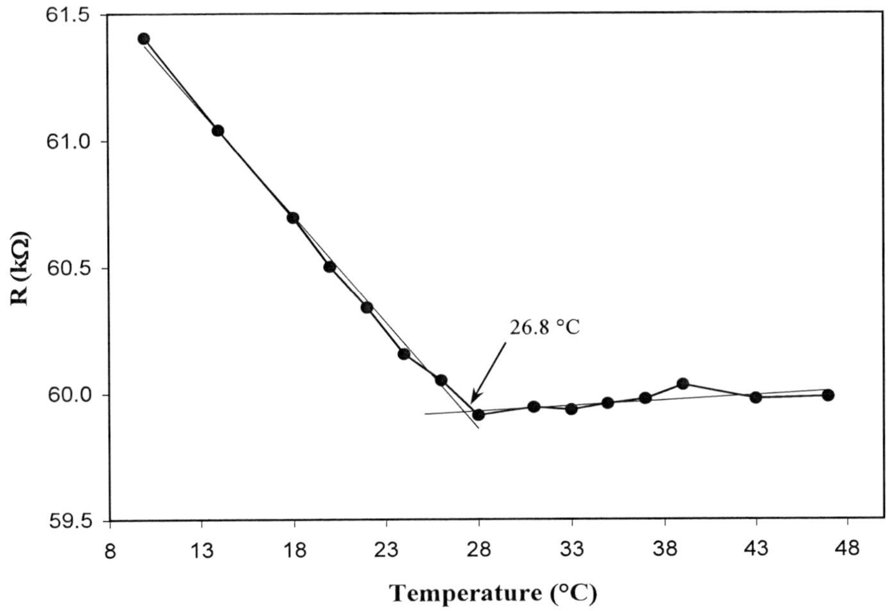

Figure 6. Resistance of [BAC-DMEDA]Co as a function of temperature. The composite materials were heated from 10 to 50°C at a rate of 0.1°C min^{-1}.

Table 2. Comparison between T_g and T at the discontinuity point

Vapor detectors	T_g [a] [°C]	T_d [b] [°C]
[BAC-DMEDA]-CB	25	24.8
[BAC-DMEDA]Cu-CB	24	23.6
[BAC-DMEDA]Co-CB	26	26.8

[a] Composite's Tg was evaluated using differential scanning calorimeter.
[b] T_d = point of discontinuity evaluated in the temperature dependence of composites resistance curve.

On the other hand, the glass transition temperature (T_g) of the PAA composite detectors measured using differential scanning calorimeter, correlated very well with the discontinuity in the temperature dependence of the resistance (Table 2). This behavior is as expected considering that the phase change between glassy region below Tg and rubbery phase above Tg, produces a change in the connectivity of the network and consequently in the electrical conduction through the conductive filler.

This property of the carbon black composites can be exploited to a simple and inexpensive method to determine the glass transition temperature of composites, representing also a good indicator of the temperature above which the detector should be operated to obtain high diffusivities for analytes.

Conclusions

The use of multifunctional polymers as the insulating component in chemiresistive vapor detectors produces significant enhancements in sensitivity to amines and carboxylic acid vapors relatively to traditional polymers used for such application. Such PAA based detectors show selectivity toward the above vapors compared to other analytes with low polarity.

Sensor diversity is achieved varying the chemical structure of monomers as well as chelating different heavy metal ions to the same carbon black polymer composite. In both cases the high overall sensitivity of the PAA based composites for specific classes of analytes was maintained. The simultaneously measurements of frequency shifts and resistance changes have demonstrated that the higher sensitivity of these systems to amines and carboxylic acid can not be attributed to the difference in polymer/gas partition coefficients. Although several different chemical factors determine the dc electrical response of this system, in our opinion changes in polymer conformation during the adsorption process play a significant role to produce the large increase in sensitivity. Moreover a discontinuity present in the temperature dependence of the materials resistance provides a simple method to determine the T_g of the carbon black composites.

References

[1] Y.S. Rao: Studies on electrical properties of polymer thick film resistors. *Microelectronics International,* 24 8, (2007).

[2] M.C.Lonergan, E.J.Severin, B.J.Doleman, S.A.Bearber, R.H.Grubbs, N.L.Lewis: Array-based vapour sensing using chemically sensitive, carbon-black-polymer resistors. *Chem. Mater.,* 8, 2298, (1996).

[3] M.J.Swann, A.Glidle, L.Cui, J.R.Baeker, J.M.Cooper: The determination of gaseous molecular density using a hybrid vapour sensor. *Chem.Commun.,* 24, 2753, (1998).

[4] S.V. Patel, M.W.Jenkins, R.C.Hughes, W.G.Yelton, A.J.Ricco: Differentiation of chemical components in a binary solvent vapour mixture using carbon/polymer composite-based chemiresistors. *Anal. Chem.,* 72, 1532, (2000).

[5] F.J. Lux: Models proposed to explain the electrical conductivity of mixture made of conductive and insulating materials. *Material Science,* 28, 285(1993).

[6] A.L. Kukla, A.S. Pavluchenko, Y.M. Shirshov, N.V. Konoshchuk, O.Y. Posudievsky: Application of sensor arrays based on thin films of conducting polymers for chemical recognition of volatile organic solvents. *Sensors and Actuators B,* 135, 541, (2009).

[7] M. Penco, L. Sartore, F. Bignotti, S. Della Sciucca, V. Ferrari, P. Crescini, S. D'Antone: Use of compatible blends to fabricate carbon black composite vapor detector. *J.Appl.Polym.Sci.,* 91, 1816, (2004).

[8] E.J Severin, N.S. Lewis: Relationship among resonant frequency changes on a coated quartz crystal microbalance, thickness changes, and resistance responses of polymer-carbon black composite chemiresistors. *Anal. Chem.,* 72, 2008, (2000).

[9] H. Lei, W.G. Pitt, L.K. McGrath, C.K. Ho: Modeling carbon black/polymer composite sensors. *Sensors and Actuators B,* 125, 396, (2007).

[10] M.D. Woodka, B.S. Brunschwig, N.S. Lewis: Use of spatiotemporal response information from sorption-based sensor arrays to identify and quantify the composition of analyte mixtures. *Langmuir,* 23, 13232, (2007).

[11] E.S. Tillman, M.E. Koscho, R.H. Grubbs, N.S. Lewis: Enhanced sensitivity to and classification of volatile carboxylic acids using array of linear poly(ethylenimine)-carbon black composite vapour detectors. *Anal. Chem.,*75, 1748, (2003).

[12] E.S. Tillman, N.S. Lewis: Mechanism of enhanced sensitivity of linear poly(ethylenimine)-carbon black composite detectors to carboxylic acid vapors. *Sensors and Actuators B,* 96, 329, (2003).

[13] T. Gao, M.D. Woodka, B.S. Brunschwig, N.S. Lewis: Chemiresistors for array-based vapor sensing using composites of carbon black with low volatility organic molecules. *Chemistry of Materials*, 18, 5193, (2006).

[14] L. Sartore, M. Penco, F. Bignotti, G. Sberveglieri: Composite materials for sensing applications and chemical sensors comprising said materials. International Patent n. PCT/IB03/01056; WO 03/078990 A1; IT TO2002A000244 (2003).

[15] L. Sartore, M. Penco, S. Della Sciucca G. Borsarini, V. Ferrari: New carbon black composite vapor detectors based on multifunctional polymers. *Sensors and Actuators B,* 111-112, 160, (2005).

[16] C.A. Mills, K.T.C. Chai, M.J. Milgrew, A. Glidle, J.M. Cooper, D.R.S. Cumming: A multiplexed impedance analyzer for characterizing polymer-coated QCM sensor arrays. *IEEE Sensors Journal,* 6, 996, (2006).

[17] P. Ferruti, E. Ranucci, F. Bignotti, L. Sartore, P. Bianciardi, M.A. Marchisio: Degradation behaviour of ionic stepwise polyaddition polymers of medical interest. *J. Biomater. Sci. Polymer Edn.,* 6, 833, (1994).

[18] L. Sartore, M. Penco, F. Bignotti, I. Peroni, M.H. Gil, M.A. Ramos, A. D'Amore: Grafting of selected presynthesized macromonomers onto various dispersion of silica particles. *J. Appl. Polym. Sci.,* 85, 1287, (2002).

[19] L. Sartore, M. Barbaglio, M. Penco, P. Bergese, E. Bontempi, P. Colombi and L.E. Depero: Polymer –coated Quartz crystal microbalance chemical sensor for heavy cations in water. *J. Nanosci. Nanotechnol.*, 9(2), 1164, (2009).

[20] L. Sartore, M. Penco, F. Bignotti, C. Pedrotti, S. D'Antone: Organic-inorganic hybrid materials with the ability to bind metal ions: calorimetric properties and thermostability. *Macromolecular Symposia,* 218, 221, (2004).

[21] M.E. Koscho, R.H. Grubbs, N.S. Lewis: Properties of vapor detector arrays formed through plasticization of Carbon black-Organic polymer composites. *Anal. Chem.,*74, 1307, (2002).

In: Chemical and Biochemical Kinetics: New Perspectives ISBN 978-1-61209-224-9
Editor: Gennady Zaikov

Chapter 10

ELECTRIC CONDUCTIVITY OF POLYMER COMPOSITES AT MECHANICAL RELAXATION

J. N. Aneli,* G. E. Zaikov[2] and O. V. Mukbaniani[1]
[1]I.Javakhishvili Tbilisi State University, Chem.Department, Tbilisi, Georgia
[2]Emanuel Institute of Biophysical Physics of RAS, Moscow, Russia

ABSTRACT

Effects of mechanical relaxations on the relaxation phenomena of electrical conducting polymer composites (ECPC) have been investigated. It is experimentally shown that filler content significantly affectsrelaxation characteristics of the material. The main reason of these differences is the effect of the filler type on proceeding of relaxational processes. Rubbers with active carbon blacksdisplay the ranges of slow and fast relaxations much more clearly, than those with low active carbon blacks even at increased concentrations in increase of the rubber elasticity modulus with the increase of the filler concentration. The effect of the filler type is displayed by formation of an inter-phase layer and is similar to the effect of the filler content. In the case of active carbon blacks, the increase of the filler-filler and polymer-filler interactions is balanced by high content of low active carbon blacks, because in both cases the modulus of the material and internal friction in relaxation processes grow.

Keywords: electrical conducting polymer composites; filler concentration; electrical conductivity; resistance; mechanical deformations; stretching; relaxation time.

1. INTRODUCTION

Mechanical stresses, induced by deformation of the composite materials, affect both the molecular and super-molecular structures of the polymers [1-3]. Structural changes occurring in composites produced by various mechanical deformations (stretching, compression, shear, etc.) affect the structure of an electrically conducting system. As a consequence, they affect the substructureand filler particles bound to it [4-6]. It is evident that in the case of conducting polymer composites mechanical deformation influences on the electrical conductivity ofmaterials.

2. Results and Discussion

It is well known that forces formed in polymer matrix from the very beginning of deformation counteracting the external influence. Therefore, the elucidation of relaxation effects on electric conductivity is complicated during deformation of ECPC due to overlapping of several factors.

To study electrically conducting properties of polymer composites during their mechanical relaxation the tests were performed, which used the same composites, studied at high deformations [7].

The experiment on determination of electrical volumetric specific resistanceρ during relaxation of the mechanical tensionσ was performed as follows. A thin sheet (2 mm) sample of electrically conducting rubber based on polydimethylmethylvinilsiloxan (industrial type – SKTV) was stretched on a stretching machine at a definite deformationε. The change of ρ was recorded by an automatic recorder immediately after stretching stop. The automatic recorder scale was graduated in Ohms on time τ, and synchronous recording of the time dependence of σ at fixed deformation was made. Moreover, relaxation characteristics were recorded for the same samples after the end the 'stretching-contraction' cycle.

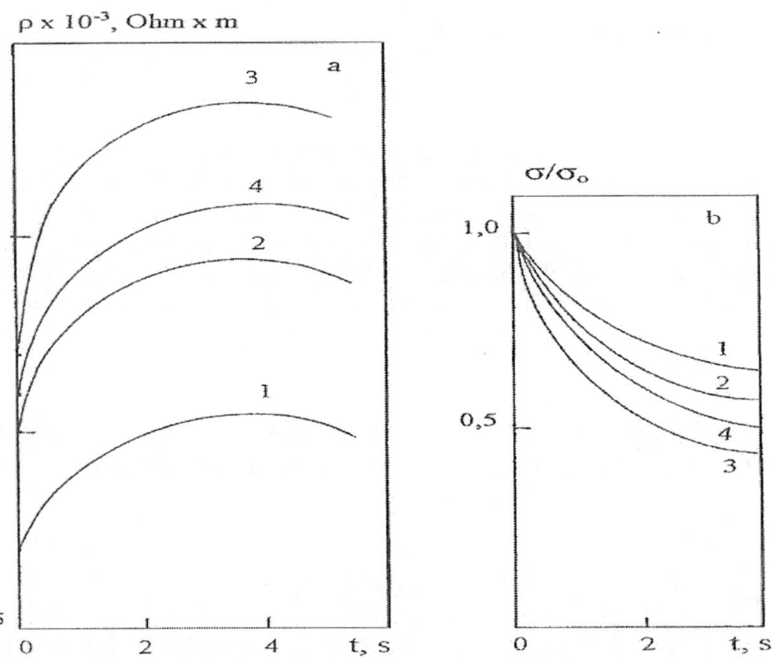

Figure 1. Time dependences of ρ (a) and σ (b) during relaxation of mechanical tension after stretching of the rubber, based on SCTV with P357E carbon black (350 mass parts) by 50 (1, 4), 100 (2), and 150% (3). The rate of elongation is 2.5 (1 - 3) and 5.0 mm/s (4), respectively.

As Figure 1a shows, the values of ρ grow during relaxation of tensionσ. In this case, the increase of ρ proceeds with higher rate in rubbers, deformed at relatively high rates. Moreover, increase of the stretching rate induces acceleration of ρ increase in these materials. In this case, kinetics of mechanical tension σ reduction is completely coincident with the character of ρ change. This means that the sample relaxes mechanically the faster, the higher the rate of the sample deformation is (Figure 1,b). Growth of ρ with time during relaxation of σ can be explained basing on the effect of the elastomer molecular system disordering that occurs after deformation release. In this case, internal mechanical tensions are reduced by transition of highly regulated system into lower regulated state on the present stage of deformation. Disordering in the system of macromolecules induces conducting circuit destruction, i.e. growth of ρ of the material. In view of the fact that the growth of the deformation rate induces increase of structural transformation intensities in the material, it is evident that significant deformations of the conducting system will happen in the electrically conducting system interacting with the polymer matrix.

The analysis of time dependences of ρ at the relaxation of rubbers show that they are successfully described by the expression:

$$\rho = \rho_\infty - (\rho_\infty - \rho_0)e^{-t/\tau}, \qquad (1)$$

where ρ_0 and ρ_∞ are border values of ρ during relaxation under fixed stretching deformation; τ is the relaxation time.

Eq. (1) represents the solution of the differential equation, similar to that [1] deduced for kinetics of the polymer mechanical relaxation. According to Eq. (1) kinetics of relaxation are described by the Maxwell model [2]:

$$\frac{d(\sigma - \sigma_\infty)}{dt} = E_1 \frac{dE}{dt} - \frac{\sigma_1 - \sigma_\infty}{\tau}.$$

Taking logarithm from Eq. (1), we get:

As Figure 2 shows, time dependences by Eq. (2) are straightened. This allows to use the model of consequently connected strings and damper (the Maxwell model) for describing kinetics of the change of ρ during relaxation of mechanical tension of electrically conducting rubber.

$$\ln \frac{\rho - \rho_\infty}{\rho - \rho_0} = -\frac{t}{\tau}. \qquad (2)$$

Thus in the case, when application of Eq. (1) to relaxation kinetics is correct, the dependence should fit a straight line in coordinates of Eq. (2).

Relaxation times τ for the composites of SCTV elastomer with P357E carbon black were determined by tangents of straight lines (Figure 2). Comparison of mechanical relaxationcharacteristics with analogical ones obtained for electrical conductivity ofsame materialsleads to conclusion that the mechanical and electrical conductivity relaxationsare in good correlation. Moreover by means of electrical conductivity there are possible to obtain in the solid polymers new relaxation phenomena invisible for other methods.

The data shown in Table 1 testify that the increase of deformation of polymers leads to the growth of the relaxation rate of both ρ and σ. This effect is strengthened at the increase of the filler concentration in the composite. For example, at comparable rates of deformation ρ of rubber containing 60 mass parts of active carbon black relaxes with higher rate than in the same rubber containing 50 mass parts of the same carbon black. In this case, the effect of the interphase growth is observed. This growth causes an increase of conditionally equilibrium modulus of the system. In composites containing lower active carbon black (P803) relaxation times are much higher than in rubbers with active carbon blacks. This point out a relatively low adhesive interaction between the polymer and the filler in rubbers with P803 carbon black. However, weak absorption forces also grow at deformation (tension increases).

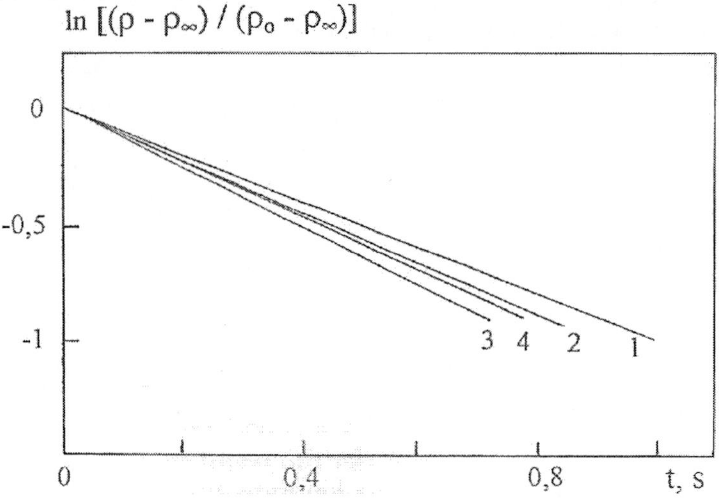

Figure 2. The dependence of ρ of the samples based on SCTV elastomer with P357E carbon black (50 mass parts) on time during relaxation in coordinates of Eq. (1). Numeration of curves coincides with one presented on Figure 1.

Table 1. Relaxation characteristics of stretched ECPC on the basis of SCTV*

Filler, mass parts	Deformation, %	Deformation rate, mm/s	Relaxation time τ, s
P357E (50)	50	2.5	1.00
	50	5.0	0.88
	100	2.5	0.91
	150	2.5	0.84
P357E (60)	30	2.5	0.60
	60	2.5	0.44
P803 (50)	100	2.5	3.84
P803 (60)	100	2.5	2.28

* Elastomers were obtained by the additive vulcanization technique.

Table 2. Relaxation characteristics of electrically conducting rubbers based on SCTV elastomer at the end of deformation cycles*

No	Filler, mass parts	Sample prehistory		Relaxation rate, s^{-1}	
		Maximal deformation in the cycle, %	Deformation rate, %	Fast	Slow
1.	P357E (40)	150	2.5	1.2×10^{-2}	7
2.	P357 (50)	150	2.5	4.0	21
3.	P357E (60)	60	2.5	6.7×10^{-2}	28
4.	P803 (60)	200	2.5	—	6
5.	P803 (120)	150	2.5	2.2×10^{-2}	11
6.	P357 (40)	150	2.5	10.6×10^{-2}	48
7.	P803 (60)	150	2.5	5.1×10^{-2}	24
8.	P803 (60) + SCTN (20)**	200	2.5	2.8×10^{-2}	16

* Composites No. 1 - 5 were obtained by the additive vulcanization technique;Composites No. 6 - 8 were obtained by the peroxide vulcanization technique.

** SCTN –low molecular (M.m. about 25 000) elastomer.

The investigation of relaxation processes at the end of the deformation cycle (stretching-contraction) after complete discharge of the samples showed that relaxation phenomena are more complex if compared with the above described facts. First of all, the complexity is expressed in the functional ρ-t dependence starting from the complete end of the cycle (Figure 3). Since the curves of the present figure were considered, let us mention that they, in fact, represent a superposition of at least two relaxation processes. One of these processes relates to the most linear parts of the curves, and the second one - to most curvilinear parts with limit overwhealming at $t \to \infty$ (see Figure 3, curves 2 and 3). Both parts of the curves mentioned reflect regulation of the conducting system by regulating the macromolecular part of the composite. However, the rate of relaxation and time of reaching equilibrium values of ρ at the end of the relaxation period significantly depend on the filler concentration. For the composite containing 40 mass parts of carbon black, the process proceeds at a low rate, but the equilibrium state in it occurs rather rapidly, whereas in rubbers with 50 and 60 mass parts of P357E higher rates are observed at the initial stage of relaxation, and decelerated curve growth - on the further stages of kinetics. One more fact is characteristic for the present dependences: the difference between minimum and maximum values of ρ on the whole interval of relaxation also strictly depends on the filling degree.

Analysis of the dependences, shown in Figure 3, gives a possibility to conclude the following. Taking into account the ideas by Bartenev /1/ on the formation of a complex heterogeneous system after injection of active carbon blacks into a polymer, which are characterized by non-linear viscoelasticity (the processes of this type are non-linear relaxation at low deformations and non-linear viscoelasticity at high deformation, the tixotropic effect by Mallins-Patrikeev [6], in particular), proceeding of two physical processes at relaxation of the considered systems can be supposed. The first process is close to the elastic range of deformation (Hooke's range) of a macromolecular system. As it is known, this process proceeds fast by both branches of the deformation cycle and is defined by elastic properties of the composite matrix. Elastic forces rapidly drive the system into the equilibrium after external influence release. However, it is often difficult to separate this process in polymers. In the composite material electrically conducting system, connected to the adsorption forces, can clearly response to any smallest structural changes in topology of structural polymer units (globules, for example), absolutely controlled by them. In the present case, conducting system in the polymer matrix plays the role of a relay-contact scheme translating information about the state of the surrounding (polymer) medium to the 'language' of electric conductivity. Thus, the abrupt decrease of ρ values, shown in Figure 3, corresponds to reduction of conducting channels existing in tight connection with macromolecules of elastomer, elastic properties of which is often promoted by the bonds mentioned.

The second process is the reduction of initial system (before the deformation) with participation of filler particles, splitted from macromolecules during deformation (stretching) of the polymer system or absent in the adsorption connection with macromolecules during deformation (carbon black agglomerates, for example). As it is known, these particles decelerate regulation in the system of macromolecules, intensify internal friction and non-linear effects at high deformations, connected to it. That is why, their existence in the composite always leads to elongation of relaxation processes. Apparently, this is the reason of deceleration of restoration of the primary conducting system of the composite material. Thus, taking into account the above mentioned it becomes clear the delay of complete reduction of the equilibrium structure in high-filled polymer matrices in higher degree comparing with low-filled polymers. For the first glance, this position seems to be paradox. Let us remind now that according to the exponential dependence of ρ on concentration of the active filler reduction of the structure is much higher than for the composite possessing 50 mass parts of the same carbon black than for the composite with 60 mass parts of the same carbon black.

The ρ - τ dependences shown on Figure 3 reflect morphological changes in heterogeneous system. Analysis of these curves showed that exponential parts are successfully described by the (1) type of equation with the only difference that ρ_∞ is substituted by ρ_m at the end of relaxation, and ρ_0 by ρ_1, corresponded to the value of ρ at crossing exponential curve by the straight line.

Using Eq. (1) for various composites tested, relaxation times were determined. Numerical values of them are shown in Table 2.

The analysis of the data from Table 2 shows that:

1. Filler content significantly affects relaxation characteristics of the material. For example, the rubber containing 40 mass parts of P357E carbon black possesses the rate of fast relaxation about 5 times lower, and duration of slow relaxation is 4-fold lower than for analogous rubber with 60 mass parts of the same filler. The rubber containing 60 mass parts of P803 carbon black has no range of fast relaxation at all, but the same rubber displays such ranges at the filling degree of 120 mass parts of P803. The main reason of these differences is in increase of the rubber elasticity modulus with the increase of the filler concentration. On the one hand, this leads to the increase of the interphase surface square, and on the other hand, to occurrence of broken polymer-filler bonds, which promote growth of internal friction and deceleration of the relaxation process in the macromolecular system. That is why, although the increase of the modulus is the reason of growth of the rate of conditionally called fast relaxation, the mechanical tensions caused by accumulation

of the filler particles, detached from polymer globules, leads to a noticeable increase of the slow relaxation time.

2. The effect of the filler type on proceeding of relaxation processes is also sufficient. Rubbers with active carbon blacks (P357E) display the ranges of slow and fast relaxations much more clearly, than those with low active carbon blacks (P803) even at increased concentrations. First of all, the effect of the filler type is displayed by formation of an interphase layer and is similar to the effect of the filler content. In the case of active carbon blacks, the increase of the filler-filler and polymer-filler interactions is balanced by high content of low active carbon blacks, because in both cases the modulus of the material and internal friction in relaxation processes grow.
3. Comparison of characteristics of the rubber, produced by two different vulcanization techniques, shows that the rate and time of relaxation depends on vulcanization network density is not weaker than the type and concentration of the filler (compare composites 1 and 6, 4 and 7, Table 2). Since the concentration of polymer cross-links in peroxide vulcanizates significantly prevails over the concentration of longitudinal links (by SCTV end groups) in rubbers, obtained by additive vulcanization, it is evident that the mechanical modulus of the first composites is higher than that of the second ones. This is the reason of differences in relaxation characteristics. In this case, the increase of the 'soft phase' in the composite by introduction of low-molecular SCTN elastomer softens relaxation processes. This is expressed by decrease of fast relaxation rate and time of slow relaxation, if compared with similar composites containing no SCTN (compare composites 7 and 8, Table 2).

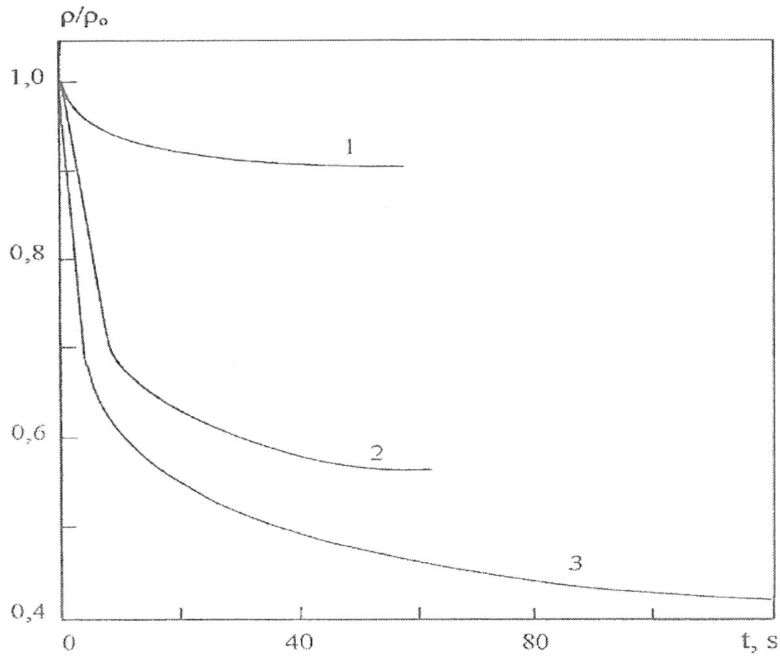

Figure 3.The dependence of ρ on delay after the contraction end (relaxation without loading)for rubbers, based on SCTV containing 40 (1), 50 (2) and 60 (3) mass parts of P357E carbon black, respectively. At the moment of discharge $\rho_0 = \rho$.

CONCLUSION

The measure of the mechanical relaxation processes in the electrical conducting polymer composites, particularly in conducting rubbers, may be investigatedby study of time-dependent processes of electrical conductance, because the behaviour of the conducting system is directly depended on any changes in polymer matrix.

REFERENCES

[1] Bartenev G.M., *Structure and relaxation properties of elastomers*.Moscow: Khimia, 1979. 288 p. (In Russian).

[2] Casale A.,Porter R. Polymer stress reactions. New-York, *Acad.* Press, 1978, 264 p.

[3] Nielsen L. *Mechanical properties of polymers and composites on their basis.* Marcell Decker Inc, New-York, 1974. 276p.

[4] Donnet A.,Voet A. *Carbon Black*, Marcell Decker, New –York- Basel, 1976. -215 p.

[5] *Carbon black polymer composites: the physics of electrically conducting composites.* Marcell Decker, New-York, 1982. 246p.

[6] Aneli J.N., Khananashvili L.M., Zaikov G.E.Structuring and conductivity of polymer composites. New-York, *Nova science publishers, 1*998. 326 p.

[7] Aneli J.N. Zaikov G.E.., Khananashvili L.M. Effect of mechanicaldeformations on conductivity of conducting polymer composites. *Journal of Applied Polymer Science,* 1999, v.74, pp. 601-621.

In: Chemical and Biochemical Kinetics: New Perspectives ISBN 978-1-61209-224-9
Editor: Gennady Zaikov

Chapter 11

ORGANOSILICON POLYMERS WITH PHOTO SWITCHABLE FRAGMENTS IN THE SIDE CHAIN

M. Doroshenko[1,3], K. Koynov[2], G. Zaikov[4], and O. Mukbaniani[1,3]

[1]Iv. Javakhishvili Tbilisi State University,
Institute of Polymer Chemistry, Georgia
[2]Max Planck Institute for Polymer Research, Germany
[3]Iv. Javakhishvili Tbilisi State University, Faculty of Exact and Natural Sciences,
Institute of Macromolecular Chemistry and Polymeric Materials, Georgia
[4]N.M. Emanuel Institute of Biochemical Physics, Moscow, Russia

ABSTRACT

Photochromism does not have a rigorous definition, but is usually used to describe compounds that undergo a reversible photochemical reaction where an absorption band in the visible part of the electromagnetic spectrum changes dramatically in strength or wavelength. In many cases, an absorbance band is present in only one form. The degree of change required for a photochemical reaction to be dubbed "photochromic" is that which appears dramatic by eye, but in essence there is no dividing line between photochromic reactions and other photochemistry. Some of the most common processes involved in photochromism are pericyclic reactions, cis-trans isomerizations, intramolecular hydrogen transfer, intramolecular group transfers, dissociation processes and electron transfers.

The photoswitchable supramolecular systems based on photochromism have been extensively investigated in recent years owing to their promising applications in photoswitching devices [1–4]. The aromatic azo group is of special interest because of the existence of cis-trans isomerism and its effect on the photochromic properties of the polymers. The great interest on polymers, containing photochromic azobenzene moiety, is driven by their potential utility for optical and photonic application and potential usage in nano- and microscale devices, such as sensors and actuators. [5]

Keywords: Photochromism, isomerization, azobenzene, hydrosilylation

PACS: Replace this text with PACS numbers; choose from this list: http://www.aip.org/pacs/index.html

In this work, we present syntheses of organosilicon oligomers with photo switchable fragment in the side chain via the hydrosilylation reaction of *α,ω-bis(trimethylsiloxy) methylhydrosiloxane* with *4-allyloxyazobenzene*in the presence of Karstedt's catalyst (1,3-divinil-1,1,3,3-tetramethyldisiloxane platinum complex {$Pt_2[VinSiOMe_2)_2O]_3$}) in a toluene solution. The catalysts used in this work were chosen on the basis of their catalytic activity in other hydrosilylation reactions and because of easy accessibility [6, 7].

4-allyloxyazobenzene was synthesized according to the following scheme:

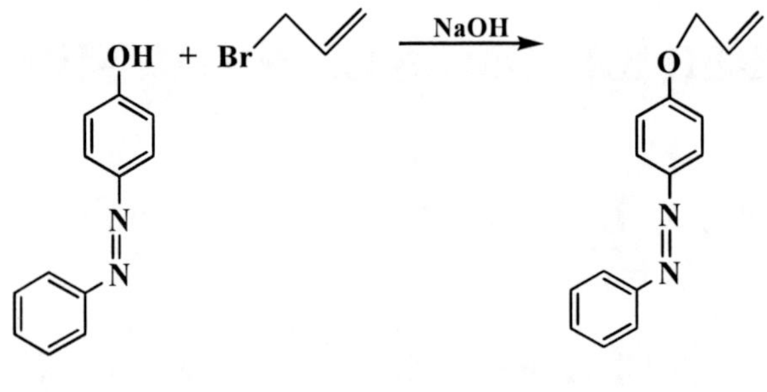

The product was purified using column chromatography. Structure and composition was determined according to the melting point, elemental analysis and ^1H NMR spectral data.

Hydrosilylation was carried out in a three-necked flask equipped with a magnetic stirrer, a tube for a catalyst inlet, a reflux condenser, and a gas inlet. The initial reaction compounds [1.07 g (0.47 mmol) of *α,ω*-bis(trimethylsiloxy)methylhydrosiloxane and 6.11 g (23.14 mmol) of *4-allyloxyazobenzene*, dissolved in 5 mL of dry toluene, were placed into the flask under an argon atmosphere and then heated until a constant temperature (60^{0}C) was obtained. Then, Karstedt's catalyst (6–8 x 10^{-5}g/ 1.0 g of starting substance) was introduced. After the completion of the reaction (14 h), the solvent was partially eliminated, the reaction product was precipitated from the toluene solution by methanol, and 6.59 g (90%) of oligomer was obtained. The progress of the conversion of active ≡Si-H groups was followed by the disappearance of the ≡Si-H absorption at 2165 cm^{-1}.

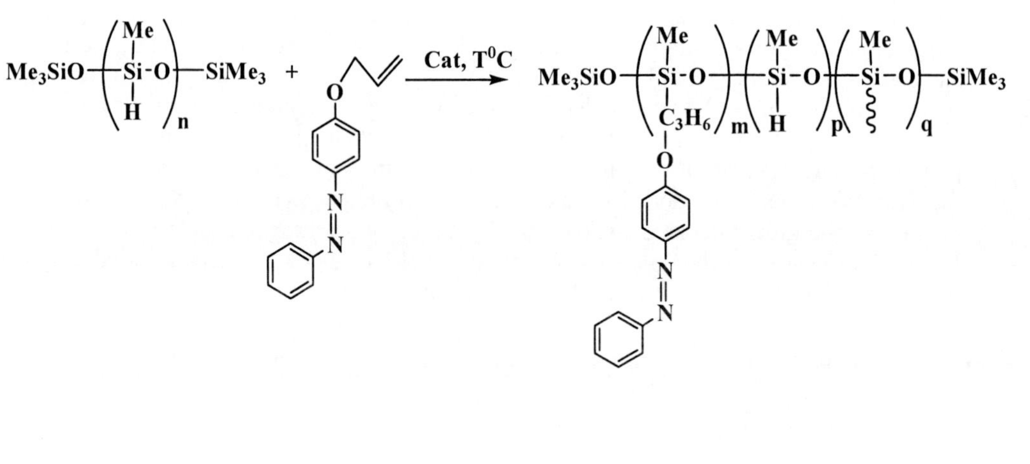

Table 1. GPC data of synthesized oligomer

Detector	Mn (g/mol)	Mw (g/mol)	D (Mw/Mn)
UV S3702	21990,10	47102,90	2,14
RI 101	21626,50	51499,70	2,66

The synthesized oligomer was a dark orange viscous liquid that was quite soluble in ordinary aromatic type organic solvents. During the syntheses of oligomer, partial gelation took place, causing the decrease in the yield. For further investigations, soluble parts were extracted from the cross linked systems. Structures and compositions of the oligomer were established on the basis of elemental analysis, FTIR and NMR spectral data. Molecular weights were determined using GPC (Table 1).

A phase behavior of the obtained oligomer was characterized with DSC. All transitions were reversible, and they did not change with repeated heating and cooling cycles.

TGA investigations were performed under a nitrogen atmosphere. It was observed that 5% mass losses for oligomer is 310°C.

It has been found by experiment that geometric isomers of azobenzene derivatives can be switched by UV-Visible irradiation. The transform can be converted to the cis form using an appropriate wavelength of light. A different wavelength of light can be used to convert the molecule back to the transform. Alternately, the molecule will thermally relax to the stable transform.

REFERENCES

[1] M. Irie, T. Fukaminato, T. Sasaki, N. Tamai, T. Kawai, *Nature* 420, 759-760 (2002).

[2] G.M. Tsivgoulis, J.-M. Lehn, *Angew. Chem.,* Int. Ed. 34, 1119 (1995).

[3] T.B. Norsten, N.R. Branda, *Adv. Mater.* 13, 347-349 (2001).

[4] T.B. Norsten, N.R. Branda, *J. Am. Chem. Soc.* 123 (8), 1784-1784 (2001).

[5] Y. Yu, M. Nakano, T. Ikeda, *Nature* 425, 145 (2003).

[6] V.P. Yuriev, I.M. Salimgareva. Hydrosilylation Reactions of Olefins, Moscow, *Nauka,* 1982, pp.123-127.

[7] H.R. Kricheldorf, *Silicon in Polymer Synthesis,* Berlin, Ed.; Springer: 1996, pp. 25-30.

In: Chemical and Biochemical Kinetics: New Perspectives
Editor: Gennady Zaikov
ISBN 978-1-61209-224-9

Chapter 12

MOISTURE PERMEABILITY AND SWELLING OF MODIFIED CELLULOSE DIACETATE FILMS

N. L. Pechnikova,* A. P. Belokyrova and T. A. Ageeva
IvanovoStateUniversity of Chemistry and Technology, Russia

ABSTRACT

The influence of chitosan and hydroxypropyl cellulose content on the process of swelling in the water and water vapor transport in mixtures with cellulose diacetate was investigated.

Keywords: cellulose diacetate, chitosan, hydrixypropyl cellulose, swelling, moisture permeability

Natural origin and wide facilities of polysaccharides chemical modification (namely cellulose, chitin and chitosan as its deacetylated derivative) attract many researches. The increased interest to biopolymers is due not only to availability and renewability of raw material but also the environmental safety of their products which are decomposed under environmental conditions [1].

To improve existed and have some new properties parent polymer is modified in a variety of ways. Polymer mixtures obtaining is an efficient method of their properties regulation.

Moisture permeability and sorptive ability of cellulose diacetate and its mixtures with chitosan and hydroxypropyl cellulose were studied in this paper.

The object of investigation was films of cellulose diacetate (CDA) and mixtures on the basis of cellulose diacetate modified with chitosan (CTS) as well as hydroxypropyl cellulose (HPC). Chitosan content in the mixture was 1%, 3%, 4,41%, 5%, 10% , 15% and that of HPC equaled 5%.[1]

Swelling kinetics of CDA and its mixtures with CTS and HPS in water at 25°C was studied. Before being tested samples were dried in the exiccator above phosphoric anhydride

* E-mail: peclin@mail.ru.
[1] Researched films were kindly afforded by Ershova U. V.

up to the constant mass at room temperature. Swelling of polymer films was investigated by gravimetric method. Preliminary studies showed that the mass of samples at drying corresponded to initial after the swelling process. Kinetic curves of all examined samples swelling are of abnormally extreme character (Figure 1). At the beginning of the process the amount of absorbed moisture ($\alpha_{мах}$) exceeds the equilibrium value ($\alpha_{равн}$). The difference between these rates ($\Delta\alpha$) is in the range 0, 88 – 2, 7%. Values of maximal and equilibrium swelling indices of cellulose diacetate mixtures with chitosan and hydroxypropyl cellulose are summarized in Table 1. The character of polymer mixtureskinetic curves at swelling doesn't depend on chitosan concentration though numerical values of swelling index change due to the increase of addition amount. The rise of chitosan content in the mixture up to 5% results in the decrease of swelling index. The increase of swelling index was observed with the increase of chitosan consentration from 10 to 15% (Figure 1). The addition of amorphous crystalline polymer(crystallinity degree of chitosan varies from 40 – 45 to 70% [2]) in the range of low concentrations results in the decrease of swelling index as crystalline regions don't take part in the process of water absorption. The influence of interphase boundary is raised due to chitosan concentration increasing in mixtures of 10% and more, the most marked structure heterogeneity appearing. At the same time the role of interface layer and microvoids also grows.

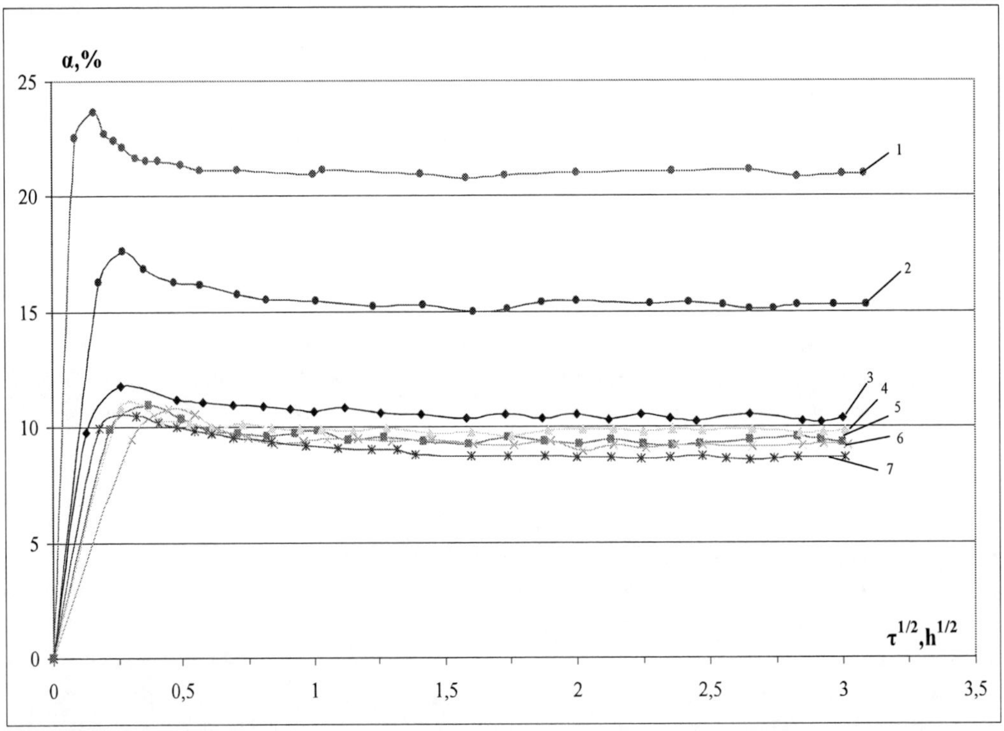

Figure 1. Swelling kinetic curves of CDA and CDA – CTS at 25°C.Chitosan content in mixture:
1 - 15%; 3 - 0%; 5 - 1%; 7 - 5%;
2 - 10%; 4 - 3%; 6 - 4,41%.

Table 1. Swelling indexes of CDA and mixtures of CDA – CTS and CDA – HPC in water at 25°C

Sample	Addition content, %	$\alpha_{мах}$, %	$\alpha_{равн}$, %	$\Delta\alpha$, %
CDA	0	11,74	10,35	1,39
CDA-CTS	1	10,92	9,35	1,57
	3	10,85	9,84	1,01
	4,41	10,77	9,17	1,6
	5	10,50	8,64	1,86
	10	17,60	15,27	2,33
	15	23,68	20,98	2,7
CDA-HPS	5	11,55	10,11	1,44

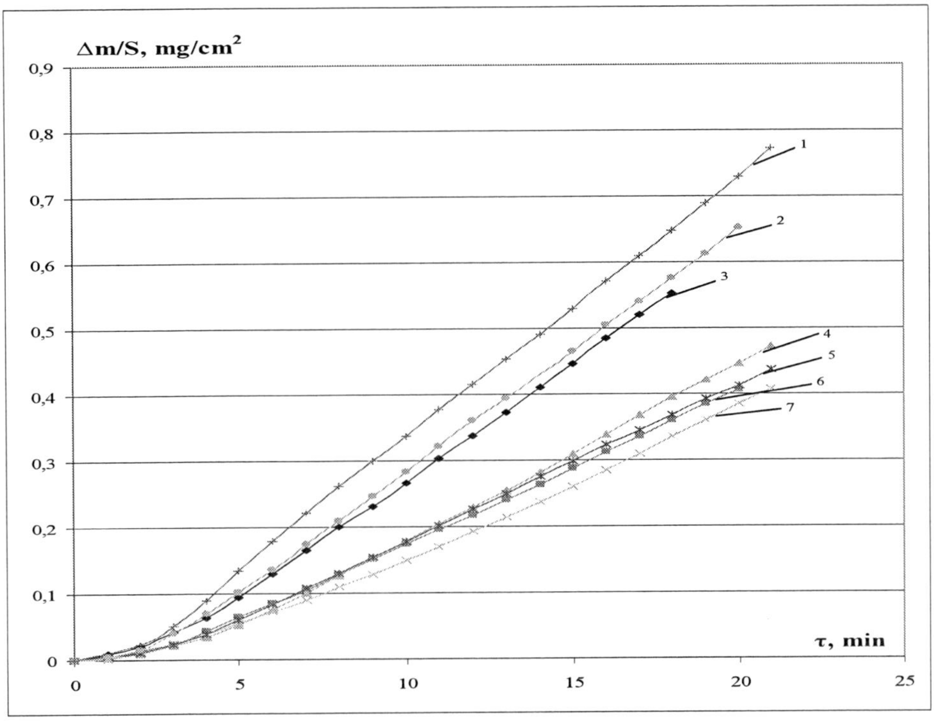

Figure 2. Moisture permeability kinetic curves of CDA and CDA – CTS at 25°C.
Chitosan content in mixture:
1 - 15% (0,0150);
2 - 10% (0,0154);
3 -0%(0,0192);
4 - 3% (0,0169);
5 - 5% (0,0140);
6 - 1% (0,0180);
7 - 4,41% (0,0176).
Film thickness is shown in brackets, mm.

Thus, not only polymers absorb water but microvoids too resulting in visible rise of equilibrium swelling index. Swelling curvers of films modified by HPC are analogous to films modified by CTS. The data in Table 1 allow estimating the influence of the second component chemical construction in the polymers mixture at their identical concentration

(5%) on swelling kinetics. According to these data sorptive ability of HPC is higher than that of CTS. It can be explained by much larger amount of OH-groups in an elementary HPC unit which have much higher hydrate values compared to the groups with chitosan. It is possible to admit that fragments $-CH_2-\underset{CH_3}{\underset{|}{CH}}-O-$ of HPC can increase its flexibility and thus raise the swelling index.

The extreme type of swelling kinetics of cellulose diacetate and its mixtures with CTS and HPC might be linked with structural change both in an individual polymer and in mixtures which initiate sorbates plastification because of retained water. As a result, crystalline region gain occurred. So the structure regulation of polymer-matrix causes maximum values increasing crystallinity degree but decreasing swelling. Our conclusions are in good agreement with the data given elsewhere [3].

Thus, summarizing the findings described above it is possible to say that kinetic curves have anomalous character provided second crystallization processes and nonequilibrium state of polymeric systems.

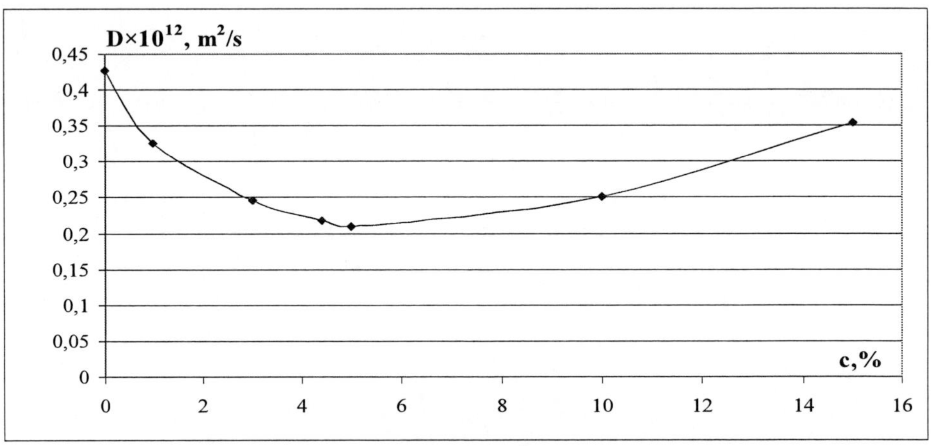

Figure 3. Dependence diffusion coefficients onchitosan consentration in the mixture.

Table 2. Permeability, diffusion and sorption coefficients of water vapors in the CDA films and their mixtures with CTS and HPS

Sample	Addition content, %	$P\times10^7$, mg×mm/ (cm^2 ×s ×tor)	$P\times10^{17}$, m^4/s×N	$D\times10^{12}$, m^2/s	$\sigma\times10^5$, m^2/N
CDA	0	4,74	2,75	0,426	6,46
CDA-CTS	1	2,96	1,89	0,326	5,80
	3	3,135	1,72	0,245	7,22
	4,41	2,92	1,61	0,218	7,39
	5	2,33	1,58	0,210	7,52
	10	3,58	2,07	0,251	8,25
	15	4,10	2,34	0,354	6,61
CDA-HPS	5	3,57	2,09	0,340	6,15

Table 3. Strength properties of CDA films andCDA – CTS and CDA - HPC mixtures

Sample	Additioncontent, %	σ_b, MPa
CDA	0	79,68
CDA-CTS	1	61,3
	3	58,77
	4,41	54,13
	5	48,75
	10	43,56
	15	8,71
CDA-HPS	5	34,28

Moisture permeability kinetics of films from cellulose diacetate and its mixtures with CTS and HPC was studied. Water vapors amount penetrating through film was controlled by the change ofquarts spiral during the time fixed by cathetometer V-630, second mixture components being CTS and HPC. Derived kinetics curves of moisture permeability of CDA and mixtures CDA– CTS at 25°C were obtained (Figure 2). Curves of CDA – HPS moisture permeability have analogous view. The amount of water vapors penetrating through films depends on the mixture composition; concentration of used addition and transmission time, the character of moisture permeability kinetic curves ofCDA, CDA – CTS and CDA – HPC films remaining constant. Values of water vapors transfer parameters in CDA, CDA – CTS with different chitosan concentration and CDA – HPC films were obtained (Table 2). In the dependence of water vapors diffusion coefficient in films of CDA and CDA– CTS polymers mixture with different chitosan content was also indicated (Figure 3). Coefficientsof moisture permeability (*P*) and diffusion (*D*) decreased with the rise of chitosan consentration in the CDA – CTS composition of polymers (Table 2). In other words, mixture permeability in theconcentrations range upto 5% is lower than initialCDA, coefficients *P* and *D* increasing in the concentrations range at 10% and more. Chitosan was said to be mainly crystalline polymer and its introduction into the polymer matrix caused the decrease of diffusion and permeability coefficients. HPC insertion into the films also drops *P* and *D* values. HPCpermeability with respect to water vapors is higher than CTS. Probably, its reasons are the same as those considered in films swelling.

Changes of efficiency coating (σ) indicate that polar groups found either in an individual state or in the mixture are available for water molecules sorption. Sorptive ability of mixturesinsignificantly differed from sorptive ability of an individual polymer. Perhaps, CTS and CDA are of nearly the same chemical structure.

Analyzed patterns were subjected to physical-mechanical tests. Breaking tension of film materials was defined, σ_b MPa values being calculated (Table 3). Breaking tension decreases for CDA – CTS mixture of polymers from 61 to 9 MPa, but for CDA – HPS it drops to 34 MPa that is lower than that of non-modified cellulose diacetate. σ_b changes due to chitosan concentration in the mixture showed that these mixtures are heterogeneous systems, besides, there is interface between the polymer matrix and included addition. Films obtained from mixtures of CDA with CTS and HPC have visually observed heterogeneity. This heterogeneity is more evidently displayed in films containing 4,41% of chitosan or more. Strength properties are considerably various because of structure inhonogeneity. It is possible to suggest that CTS insertion causes decrease of films breaking tension. Low strength of films can be explained not only by their structure inhomogeneity but also availability of macrodefects such as local thinning. Some authors [4] who studied deformative-strength characteristics of cellulose – chitosan mixture with different chitosan concentration in the mixture observed analogous dependence.

One can say that CDA – CTS and CDA – HPC mixture are incompatible according to the data analyzed (change of equilibrium swelling index, permeability and diffusion coefficients against CDA – CTS and CDA – HPC composite films). It also confirms the absence of linear dependence between α, P, D and chitosan amount in the mixture. To qualitatively estimate the compatibility of the polymer mixture can be easily done visually. CDA – CTS and CDA – HPC films are observed to be cloudy in the range of concentrations applied in contrast to parent CDA. This also indicates heterogeneity of polymer mixtures studied.

REFERENCES

[1] Tukova I. S., Syvorova A. I., Petrova A. P., Vihoreva G. A. // Vysokomol. *Soedin. Ser. A,* 2003, 45 (5), 791 – 796.

[2] Mogilevskaya E. L., Akopova T. A., et. al. // *Vysokomol. Soedin. Ser.* A, 2006, 48 (2), 216 – 226.

[3] Belokyrova A. P., Byrmistrov V. A., Sherbina A. A., Rosin M. V., Gerasimov V. K., Chalih A. E. // Izv. Vyssh. Ucheb. Zaved., *Khim. Khim. Tekhnol.* 2004, 49 (6), 42 – 45.

[4] Rogovina S. Z., Golova L. K., Borodina O. E., Vihoreva G. A. // *Khim. Volokna.* 2002, 1, 18 – 20.

In: Chemical and Biochemical Kinetics: New Perspectives ISBN 978-1-61209-224-9
Editor: Gennady Zaikov

Chapter 13

FERROCENE-CONTAINING POLYPHENYLENES AS PRECURSORS FOR MAGNETIC NANOMATERIALS

R. A. Dvorikova,[1] Yu. V. Korshak,[2] L. N. Nikitin,[1] M. I. Buzin,[1] V. A. Shanditsev,[1] Z. S. Klemenkova,[1] A. L.Rusanov,[1] A. R. Khokhlov,[1] A. Lappas,[3] and A. Kostopoulou[3]

[1]Institution of Russian Academy of Sciences A. Nesmeyanov Institute of Organoelement Cjmpounds, RAS, Moscow, Russian Federation

[2]D. Mendeleev University for Chemical Technology of Russia, Moscow, Russian Federation

[3]Institute of Electronic Structure and Laser (IESL) Foundation for Research and Technology - Hellas (FORTH) Heraklion, Crete

ABSTRACT

New magnetic nanomaterials have been synthesized from ferrocene-containing polyphenylenes. Cyclotrimerization of 1,1′-diacetylferrocene by condensation reaction catalyzed by p-toluenesulfonic acid in the presence of triethyl orthoformiate both in solution and supercritical carbon dioxide in the temperature rangeof 70 – 200°C is described. Highly branched ferrocene-containing polyphenylenes prepared by this procedure were used as precursors for preparing magnetic nanomaterials. This was achieved by thermal treatment of polyphenylenes in the range of 200 – 750°C. The emerging of crystal magnetite nanoparticles of magnetite with the average size of 6 to 22 nm distributed in polyconjugated carbonized matrix was observed due to crosslinking and thermal degradation of polyphenylene prepolymers. Saturation magnetization of such materials came up to 32 $Gs \cdot cm^3/g$ in a filed of 2.5 kOe.

Keywords: ferocene, polyphenylenes, magnetic nanomaterials, synthesis, properties

INTRODUCTION

In the last few years, the enhanced interest to nanosized materials has been developed due to potentially unusual physical properties of those items compared to common substances. The magnetic properties of such substances are of great interest and presently under intense investigations [1-9].

The magnetically active materials are known to use for many purposes including data recording and information storage, for permanent magnets, in magnetic cooling systems, as magnetic sensors etc. [1, 2]. The precise drug delivery by using magnetically active nanoparticles is of current interest in biomedical field [10] as well as new probabilities may be created in the field of MR-imaging, cell sorting and cell separation, in bio-selection processes, for enzyme immobilizing, in immune analysis, in catalytic processes, and the like [11-14].

Stabilization of nanosized magnetic particles is a key problem and may be solved by their incorporation in oligomer or polymer matrixes – one of the possible ways applicable for preparation of new magnetic materials [1, 2].

Previously, we have developed a new approach for nanosized composite preparation by thermal treatment of ferrocene-containing polymers with reactive terminal groups [15]. Thermal treatment of ferrocene-containing polymers prepared from 1,1′-diacetylferrocene in the range of 150 – 350°C was accompanied by crosslinking and origination of iron-containing nanosized particles displaying magnetic-ordering properties. The synthesis of highly-branched ferrocene-containing polyphenylenes was performed also by cyclotrimerization of 1,1′-diacetylferrocene in environmentally friendly solvent – liquid and supercritical carbon dioxide [16]. This medium had been much used as a "green environment" for conducting of various chemical processes including synthesis and modification reactions of polymers [17-24]. The new approaches for creation of metal-polymeric systems and magneto-active nanocomposites, in particular, are of special interest [25-33].

Paper [16] describes using of p-toluenesulfonic acid (p-TSA) or $SiCl_4/C_2H_5OH$ mixture as catalysts for the reaction. The reaction was performed in liquid and supercritical carbon dioxide in the presence of triethyl orthoformiate at pressure of 20 MPa and temperature of 20 and 50°C with the polymer yield around 20%. DTG-analysis revealed 5% weight loss of the polymer at 400°C with the weight of carbonized residue of 80% after heating at 750°C. When heated at 300°C, the process of crosslinking in the samples occurred and crystalline iron-containing nanosized magnetic particles with the average size of 10 – 40 nm with saturation magnetization of 13 $Gs{\cdot}cm^3/g$ in a magnetic field of 2.5 kOe were observed in a polymer matrix.

In the present paper, the study of cyclotrimerization reaction of 1,1′-diacetylferrocene has been carried out in solution and in supercritical carbon dioxide in a wide temperature range for obtaining higher polymer yields. The controlled thermal treatment of prepared ferrocene-containing polyphenylenes has been carried out for attaining improved magnetic properties.

Table 1. Synthesis of ferrocene-containing polymers (FP) with p-TSA as a catalyst and specific magnetization values of the samples

Sample	Temperature, °C	Duration/ Postreaction time at ambient temperature, hrs	Catalyst, %	Yield, %	Elemental analysis, %*				σ_H,max, G·cm^3/g (at temperature)
					C	H	Fe	S	
FP-1	70	2.5/40	10	24	66.51	4.86	20.69	0.80	9 (500°C)
FP-2	70	5/20	10	25	65.64	4.91	20.02	0.62	10 (500°C)
FP-3	100	2	10	16	64.43	4.56	12.83	2.11	21 (750°C)
FP- 4	100	2	20	24	61.51	4.34	13.10	3.18	21 (500°C)
FP- 5	100	2/15	10	45	68.81	5.45	18.55	-	12 (500°C)
FP-6	100	2/20	10	66	62.85	5.05	19.37	0.65	26 (500°C)
FP-7**	110	2/40	10	18	64.22	4.21	14.88	1.65	14 (500°C)
FP-8	120	2	10	19	64.95	4.73	12.81	2.11	18 (1000°C)
FP-9**	130	2/20	10	19	50.14	4.15	26.67	2.00	21 (750°C)
FP-10	140	2/20	20	71	67.32	5.51	20.2	-	32 (750°C)

*Calculated for $C_{126}H_{102}Fe_9O_6$: C – 66.88%, H – 4.54%, Fe - 22.21%.
**Reaction was carried out in ditolylmethane; Molar ratio TEOF/Acetyl group was 1.2.

EXPERIMENTAL

Ferrocene-containing polymers (FP) were prepared either by conventional solution method [15] or in supercritical carbon dioxide (SC-CO_2) (FPSC) using technique described in [16] at 20 MPa and temperatures varied from 70 to 200°C.

Conventional polymerization of the monomer was carried out in a flat-bottom flask supplied with a thermometer and magnetic stirrer. 1.0g (3.7 mmol) of 1,1′-diacetylferrocene, 3 ml (8.0 mmol) of triethyl orthoformiate, and 0.10gp-toluenesulfonic acid were placed into the flask and allowed to react at 70°C during 2.5 hrs and then left for 40 hrs at room temperature. The dark brown precipitate was filtered, washed with water to neutral reaction, then with ethanol and dried in vacuum. The yield was 0.34g (24% from theory). Increasing of the temperature up to 140°C raised yields to 71%. Polymerization in SC-CO_2 was performed in a high pressure reactor with the inner capacity of 10 ml. The reactor was flushed out with CO_2 after loading of the reagents and heated up to the required temperature (±0.5°C). The input of CO_2 and applying of the required pressure (20 MPa) was completed by hand-operating press ("High Pressure Equipment", USA). The reaction mixture was agitated with magnetic stirrer and the reaction was carried out from 2.5 to 5 hrs. After completing of the reaction the autoclave was cooled down and the pressure released. The resulting polymer was washed successively with ethanol and water to pH = 6 –7 and then dried. The maximum yield was almost quantitative.

Magnetoactive materials were prepared by heating of the original samples of FP and FPSC in quartz tubes placed into measurement cell of magnetometer in the range of 200–750°C, or by thermal treatment of the samples in glass tubes under argon in the range of 250–500°C.

Appearance of the magnetic order during thermal treatment of the polymers was monitored by vibration magnetometer of the Foner type.

Morphology of the nanocomposites was studied by TEM using LEO 912AB OMEGA instrument. Size distribution of nanoparticles was figured out by statistical treatment of 50–100 particle size.

X-ray powder diffraction was obtained with Rigaku D/MAX-2000H rotating anode diffractometer with CuKα radiation and secondary pyrolytic graphite monochromator. DC magnetic susceptibility measurements were carried out with an Oxford Instruments MagLab EXA extraction magnetometer.

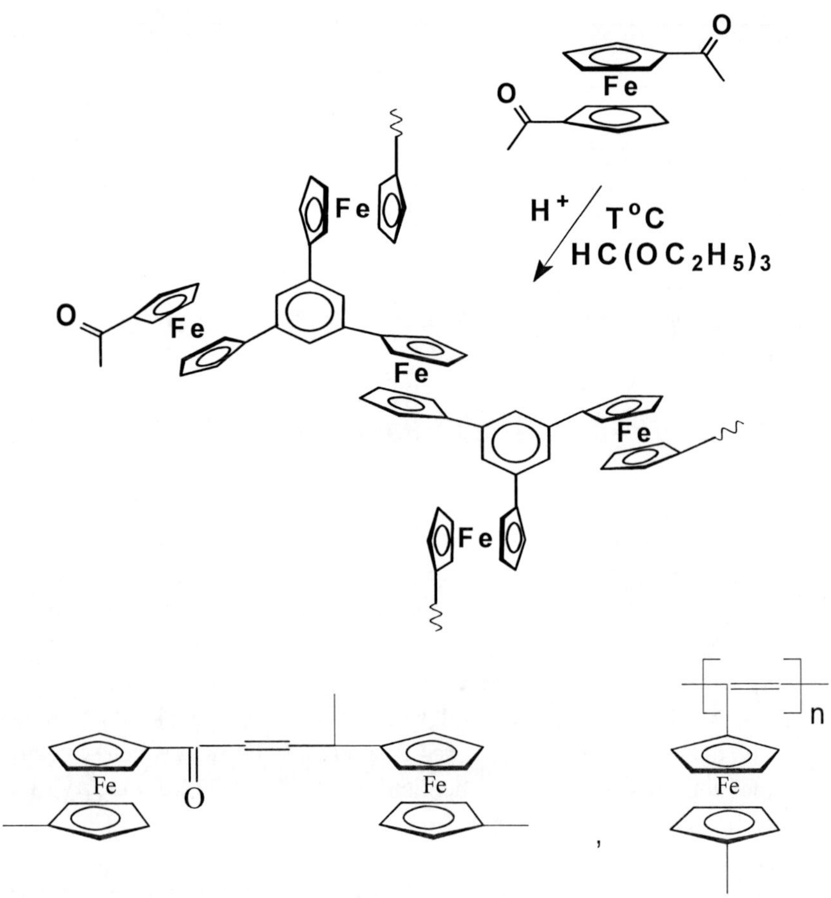

Scheme 1.

Table 2. Preparation of FPSC polymers in supercritical CO_2 and specific magnetization values of the samples after heat treatment at 500°C

Sam-ple	Temperature, ^0C	Duration/Postre action time at ambient temperature, Hrs	Catalyst, %	Yield, %	Elemental analysis, %				Magnetization ($\sigma_{H,max}$) $G{\cdot}cm^3/g$,
					C	H	Fe	S	
FPSC-1	70	2.5/20	10	32	67.43	5.17	20.76	0.25	6.0
FPSC -2	140	2/20	10	31	58.98	4.40	16.47	2.73	15.0
FPSC -3	140	2/20	60	98	53.67	4.68	23,36	0	12.5
FPSC -4	160	2/20	10	37	62.09	4.44	16.04	2.01	12.5
FPSC -5	200	2	10	92	63.43	4.95	16.81	0.80	14.5

TG analysis was carried out with "DERIVATOGRAPH-C" (MOM, Hungary) instrument in air and argon by using samples of ~15 mg at a heating rate 5 ^oC/min.

Results and Discussion

The cyclotrimerization reaction of 1,1′-diacetylferrocene catalyzed by p-TSAwas studied for determining the optimal conditions for higher yields of polyphenylenes as precursors for materials with improved magnetic properties.

Ferrocene-containing polymers were synthesized both under conventional conditions and in supercritical carbon dioxide (SC–CO_2), p-TSA was used as a catalyst for cyclotrimerization of 1,1′-diacetylferrocene. The reaction proceeded in accordance with the Scheme 1 [16] which did not consider side reactions that could bring about some defect units of dypnone (β-methylchalcone) structures [34].

The effect of reaction conditions, such as temperature, duration, catalyst concentration, and post-stirring period at ambient temperature on yield and properties of the final products was studied. Tables 1 and 2 present some data on the reaction conditions and some properties of the polyphenylenes prepared under conventional conditions (FP) and in supercritical carbon dioxide (FPSC). The reaction was carried out in the temperature range of 70-200°C in the presence of triethyl orthoformate, which acts simultaneously as a ketalizing agent and as a solvent. Besides, ditolyl methane was used in some experiments as a solvent. The resulting ferrocene-containing polyphenylenes were obtained as powders of dark-brown color partially soluble in organic solvents such as dioxane, methylene chloride and benzene.

The yields of polymers increased at 100°C from 24 to 66%. Elemental analysis showed good agreement between found and calculated data and, in addition, sulfur was found in the samples in ~2% amount. The presence of the latter may be due to the chemical interaction between polymer and p-NSA that may also explain the reduced content of iron and carbon in analysis data. The maximum yield of FP was achieved at 100°C and 10% concentration of a catalyst, whereas 20% catalyst concentration provided the highest yields at 140°C.

The increase of the reaction time at 100°C resulted in an enhancing yield from 16 to 66% (Table 1). The same effect was observed when the catalyst concentration was raised up from 10 to 20%.

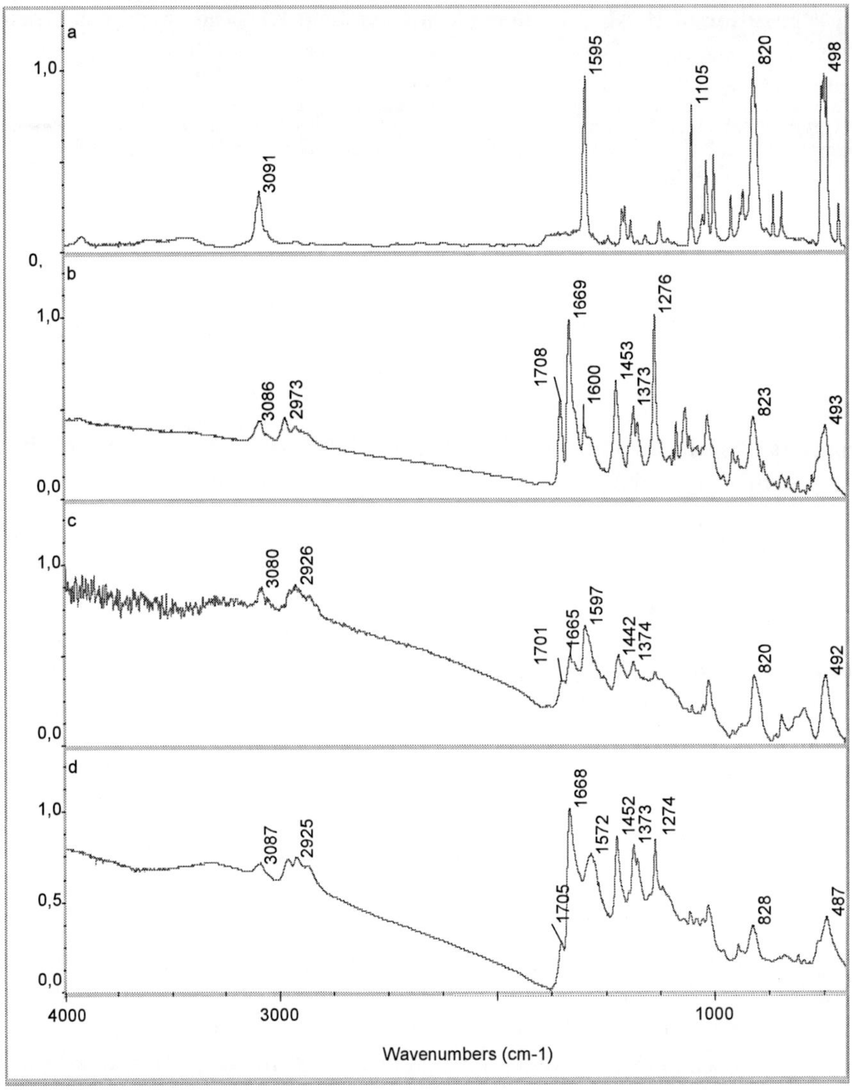

Figure 1.The IR-spectra of: a) 1,3,5-triferrocenylbenzene; b) FP-10 (table 1), c) FP-10 after heating at 350^{0}C for 1 hour in measurement cell of magnetometer; d) FPSC-5 (table 2).

Preparation of ferrocene-containing polymers in SC-CO_2 medium (FPSC) was of special interest. Some data are given in Table 2. Thus, the increase in reaction temperature from 70 to 200°C rose up the polymer yield from 10 to 90%, whereas the increasing the catalyst concentration from 10 to 60% brought about the quantitative yield of polyphenylenes.

IR-spectra of FP revealed the band at 1600 см$^{-1}$ typical for stretching modes of CH-groups in 1, 3, 5-substituted benzene rings (Figure 1b), while this band was observed at 1572 см$^{-1}$ for FPSC (Fig 1d). The bands typical for ferrocenyl units were also found IR-spectra of the polymers (stretching modes of CH-groups at 3090 см$^{-1}$, non-planar bending vibrations of CH-groups in C_p rings at 820-830 см$^{-1}$, and the bands for doubly degenerated antisymmetric stretching modes Fe-C_p at 484 см$^{-1}$). Strong bands at 1275 см$^{-1}$ are due to symmetric stretching modes of C–C in disubstituted C_p rings and intensive bands at 1670 см$^{-1}$ и 1700

$см^{-1}$ may be referred to C=O stretching vibrations in dypnonefragments and terminal acetyl groups. IR-spectrum of model 1, 3, 5-triferrocenyl benzene shows the same characteristic bands typical for the polymer samples (Figure 1a) except visible broadening of the latter due to higher molecular weight.

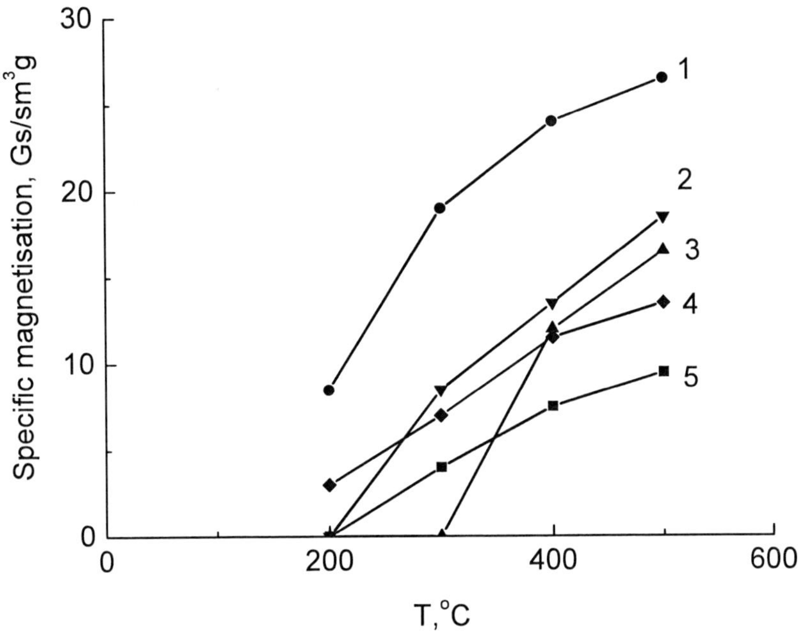

Figure 2. Magnetization curves for FPs obtained at different temperature: 1) FP-6; 2) FP-9; 3) FP-7; 4) FP-8; 5) FP-1.

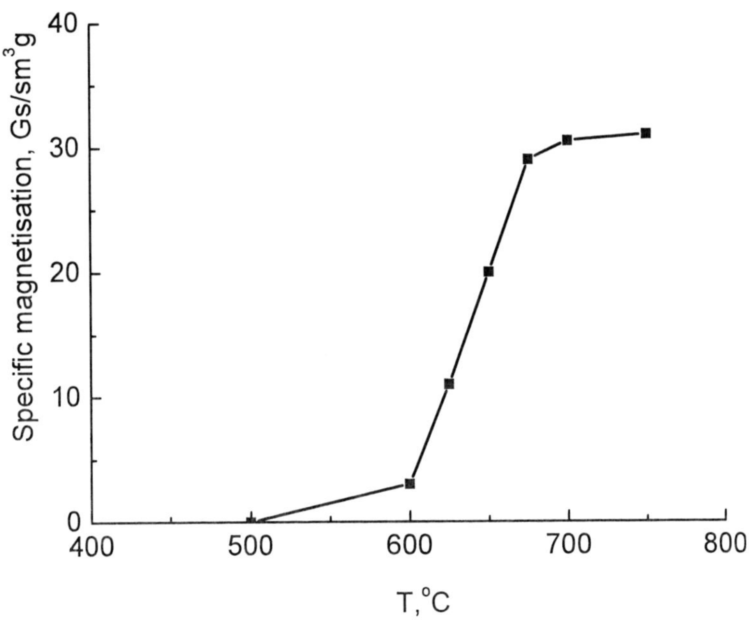

Figure 3. Magnetization curve for FP-10.

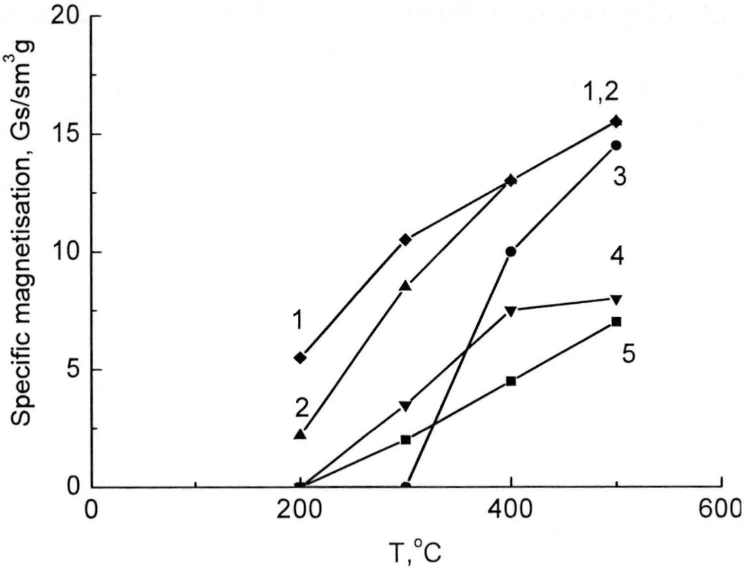

Figure 4. Magnetization curves for FPSCs obtained at different temperatures: 1) FPSC-2; 2) FPSC-4; 3) FPSC-5; 4) FPSC-3; 5) FPSC-1.

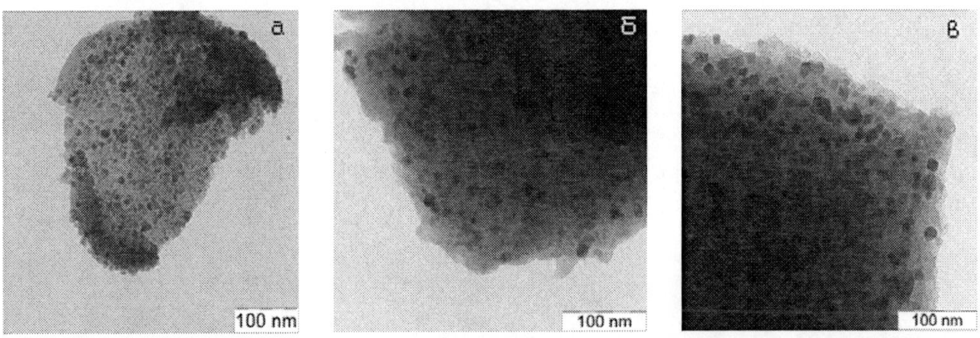

Figure 5. Microphotographs of: a) FPSC-5 after heating in argon at 250°C for 2 hours, average size of nanoparticles – 8 nm; b) FP-8 after heating in argon at 350°C for 2 hours, average size of nanoparticles – 13 nm and c) FP-6 after heating at 500°C for 1 hour (heated in measurement cell of magnetometer), average size of nanoparticles – 22 nm.

The amount of the β-methylchalcone fragments and acetyl end groups in polymers was found to reduce after heating samples at 250°C and higher temperatures as evidenced by decreasing intensities of the bands at 1600 $см^{-1}$ and 1700 $см^{-1}$. Simultaneous increase of the regular units in polymer molecules was found due to presence of 1, 3, 5-substituted benzene rings as evidenced by increasing of the band at 1600 $см^{-1}$ specific for those groups (Figure 1c). At the same time, network formation occurred and arising of the crystal iron-containing particles was observed. The mechanism of the particle formation comprises their spontaneous adjustment at the nanosized level depending on the polymer structure. Magnetic behavior of the final products is markedly influenced both by temperature regimes during ferrocene-containing polyphenylenes synthesis and subsequent thermal treatment. Thus, formation of magnetic phase in sample FP-6 (Figure 2) prepared by conventional method at 100°C (FP) begins after 200°C and the maximum magnetization is equal to 26 Gs·cm^3/g. Meanwhile, the sample FP-10 that was synthesized at 140°C showed the arising of magnetic phase after heating at 500°C, and at 750°C the magnetization approached to its maximum value of 32 Gs·cm^3/g (Figure 3).

Figure 4 presents the plot of magnetization values versus heat treatment temperature for ferrocene-containing polyphenylenes synthesized in the range of 70–200°C in supercritical carbon dioxide (FPSC). For such samples formation of magnetic phase stats at 200°C for samples FPSC-2 and FPSC-4 prepared at 140 and 160°C, respectively. The highest magnetization (~15 Gs·cm^3/g) was achieved for samples FPSC-2, FPSC-4 and FPSC-5, which were synthesized at 140,160, and 200°C, respectively (Table 2). However, FPSC-5 shows magnetic behavior just at 300°C. The lowest magnetization value after heat treatment was recorded for sample FPSC-1 originally prepared at 70°C. Besides, the samples synthesized in SC-CO_2 medium had the reduced magnetic properties compared to FP that could be explained by their higher thermostability.

TEM study of thermally treated ferrocene-containing polyphenylenes revealed in all samples presence of iron nanoparticles homogeneously distributed in polymer matrix with the average size of 6 to 22 nm depending on the synthesis conditions and heating regime. It should be mentioned that by increasing temperature of thermal treatment from 250 to 500°C one could increase the size of nanoparticles from 8 to 22 nm for the samples obtained in the presence of p-TSA as a catalyst. TEM images are shown in Figure 5a and 5b for samples FPSC-5 and FP-8 subjected to heating at 250 and 350°C, respectively, in argon for 2 hrs. The size of nanoparticles increased when the temperature of heat treatment rose up. Figure 5c shows TEM image revealing nanoparticles with the average size of 22 nm for FP-6 sample after heating at 500°C in argon for 1 hour. In the meantime, the particle size in a sample prepared with 20% of a catalyst and treated at the same temperature of 500°C was half as much and equal to 10,6 nm (Figure 6a) and the subsequent heating up at 700°C resulted in reducing to 6.3 nm (Table 3, Figure 6b).

Table 3 presents the data on nanoparticle size distribution for FP-7 sample after heat treatment at 500, 600, 650, 675, and 700°C. It is evident that increasing of temperature brought about the visible effect on the decreasing of the mean size of nanoparticles. Moreover, the particle size distribution for the sample was rather narrow when it was kept at 500 °C compared to higher temperatures.

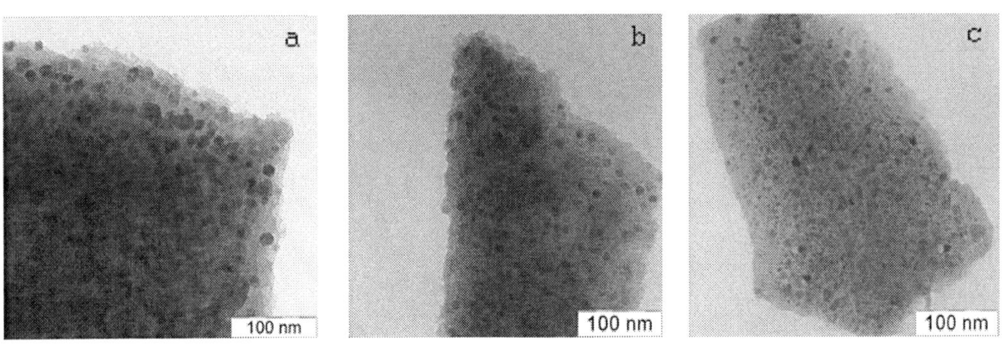

Figure 6. Microphotographs for FP-6 (a) and FP-10 (b, c), heated in measurement cell of magnetometer for 1 hour at 500 (a, b) and 700°C (c). The average size of particles – 22 (a), 10.63 (b) and 6.32 nm (c).

Table 3. Size distribution of magnetic nanoparticles as a function of heat treatment temperature

Heat trearment temperature,°C	500	600	650	675	700
Average nanoparticle size, nm	10.63	9.21	6.96	6.66	6.32
Maximum size of nanoparticles, nm	13.63	15.21	25.23	22.07	20.56
Minimum size of nanoparticles, nm	8.55	4.17	2.35	2.09	2.14

Thermal and thermooxidative stability of synthesized ferrocene-containing polymers was studied by TGA method. It was found that thermal and thermooxidative degradation followed the basic features of such processes. Thus, the higher degree of decomposition in air was observed for all investigated polymers compared to that in an inert atmosphere. As an example, Figure 7 shows TGA curves for FP-10 sample. The weight loss of the sample in air was about 70% whereas in inert atmosphere the sample loses essentially less from the initial weight (about 30%). The weight loss, both in air and in argon was practically coinciding and begins at the same temperature (190°C). Then, the thermal decomposition under heating in air was slowed down, and the sample starts to lose weight intensively only in the vicinity of 400°C. Such a behavior can be attributed to a formation of the dense network in the polymeric matrix at the initial stage of the thermooxidative degradation that prevents thermal degradation for a while. However, the peroxides formed during the polymer oxidation in air may form a redox system together with ferrous species in ferrocene units to yield additional amounts of free radicals that finally stimulate a deeper degradation of the polymer. On the contrary, in an inert atmosphere the sample slowly loses weight in a wide temperature interval, up to 700°C and then its weight does not change.

The most significant observation concerns the comparison of the XRD spectra for the samples prepared by conventional way and in SC-CO_2 (FP-6 and FPSK-5) and heated at 250°C, and the models shown in Figure 8c. We may conclude from Figure 8a-8c that the only crystalline phase presents in the samples if that of magnetite Fe_3O_4 and there is an excellent match between the experimental data and Fe_3O_4 model.

In addition to this crystalline phase there is a significant volume fraction arising from an amorphous phase, which is identified by the diffuse scattering (broad hump) in the 2Θ range 10-20 degrees. Based on the excess broadening of the Bragg reflections of Fe_3O_4, in sample FPSC-5, we suggest that the magnetite is a little less crystalline than that present in FP-6. Probably this is due to the different conditions of the synthesis for this material, since the sample was prepared in SC-CO_2 medium.

To this extent the following results, on the magnetic properties of these samples, are mainly dictated by the behavior of the magnetite which is magnetic (ferromagnetic) already at room temperature (Figure 9). Although the measured magnetization, at T~ 280 K, is characteristic of a soft ferromagnetic-like material, with little coercivity (H_c a few Gauss), there are a some qualitative differences (e.g. steep change of M vs. H for FP-6) between the curves of the two materials, probably due to the higher sample crystallinity (FP-6) or/and larger particle size. As far as the magnetic susceptibility is concerned (Figure 10), we find an important reduction of the magnetic moment at relatively low temperature ($T<T_B$) for both samples. Furthermore, FP-6 displays a second, higher "blocking" temperature (T_{B2}~ 100 K) which may be related to the Verwey transition (~118 K) met in bulk magnetite – shifted to relatively lower-T here maybe due to smaller particle-size.

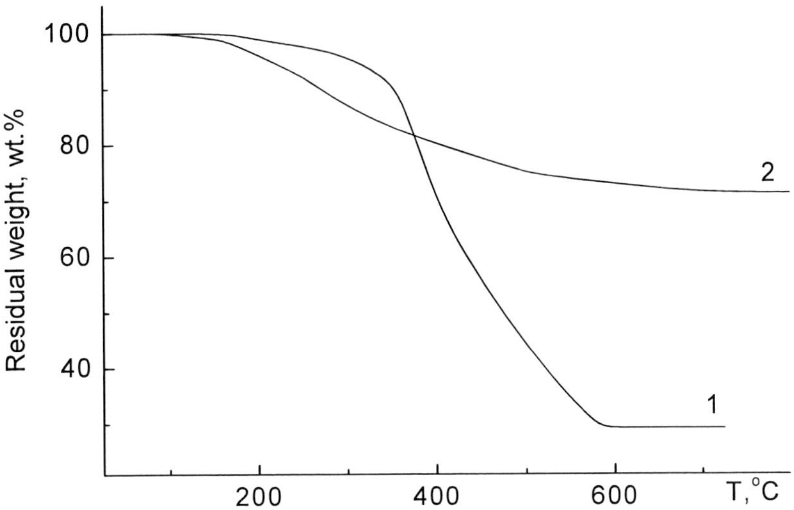

Figure 7. TGA curves for FP-10 in air (1) and in argon (2) at a heating rate 5°C/min.

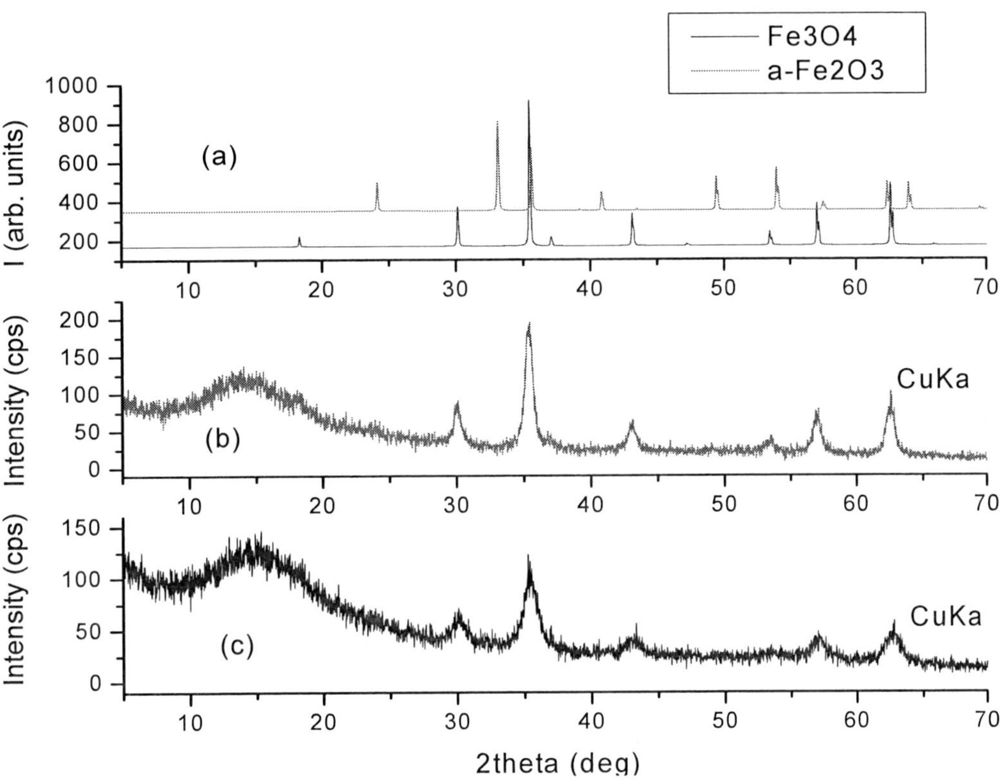

Figure 8. X-ray powder diffraction (XRD) patterns of two ferrocene-containing polyphenylene samples FP-6 (b), and FPSC-5 (c) after heating at 250°C, and the model spectra of two iron-based oxides (a).

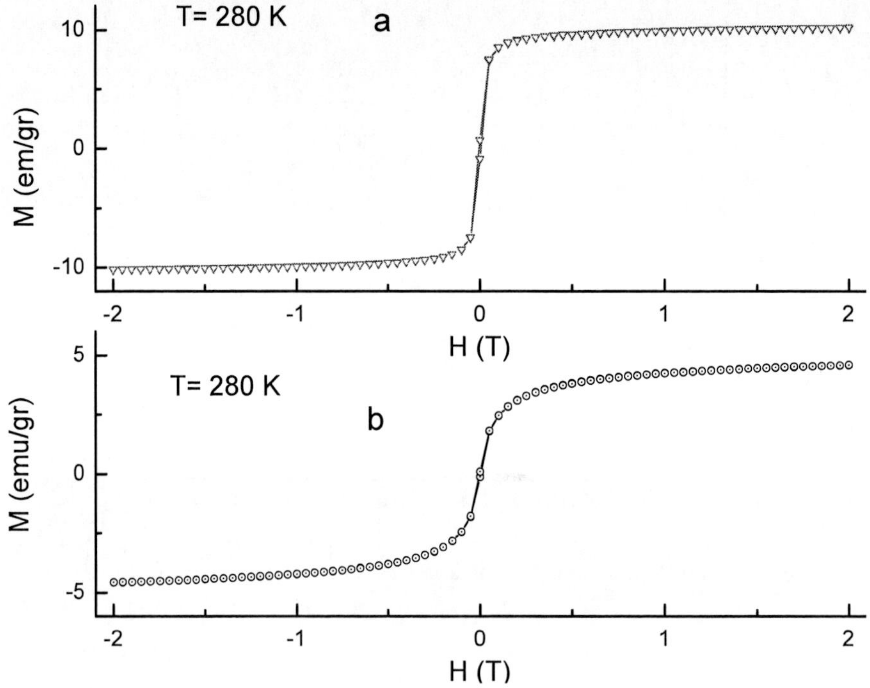

Figure 9. Magnetization curves for FP-6 (a) and FPSC-5 (b) at room temperature.

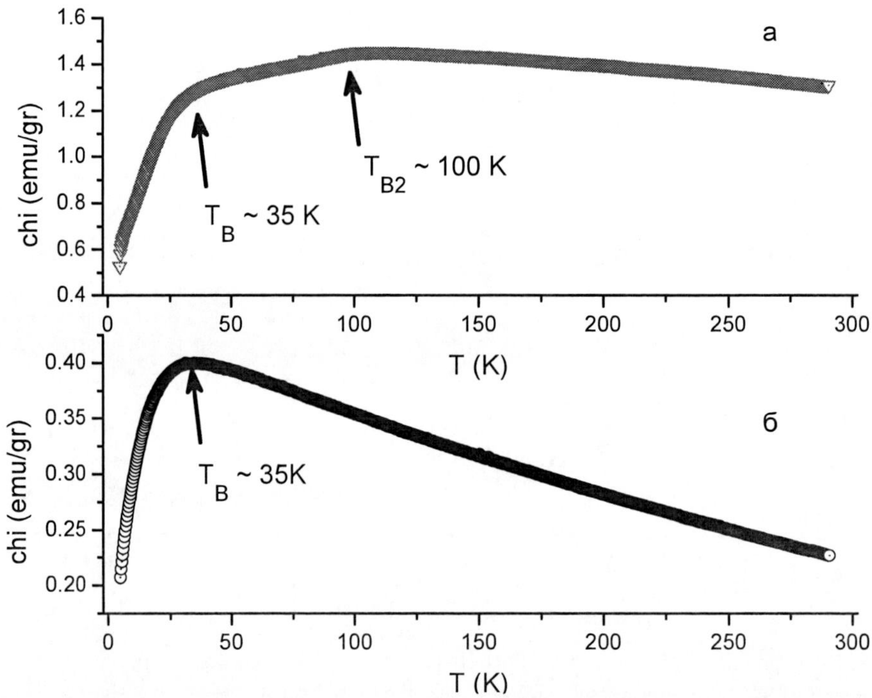

Figure 10. Temperature dependence of the zero-field cooled (ZFC) magnetic susceptibility of FP-6 (a) and FPSK-5 (b). The arrows indicate various "blocking" temperatures (T_B) in these samples.

SUMMARY

The optimal conditions for synthesis of highly branched ferrocene-containing polyphenylenes were developed via cyclotrimerization of 1,1′-diacetylferrocene by condensation reaction in the range of 70 – 140°C providing 71% yield of the polymer. The quantitative yields of the polymers were obtained in supercritical carbon dioxide (SC-CO_2).

Magnetic nanocomposites with saturation magnetization of 32 $Gs \cdot cm^3/g$ in a field of 2.5 kOe were prepared by thermal treatment of such polymers in the range of 250 – 750°C. TGA data approved that initial temperature of sample degradation fit in with magnetic phase formation. XRD studies and magnetic measurements showed that crystalline phase of magnetite Fe_3O_4 presents in the samples with magnetization characteristic of a soft ferromagnetic-like material with small coercivity.

TEM study of polymer samples approved that the mean size of magnetic nanoparticles in polymer matrixes may be controlled by varying reaction conditions and heat treatment regime of the final polymers.

ACKNOWLEDGMENT

We are thankful to S.S. Abramchuk for transition electron microscopy imaging (TEM), and to Dr. S.P. Solodovnikov for helpful discussions. We are grateful to the Russian Foundation for Basic Research (RFBR) for funding these research projects 08-03-00294, 09-03-91227, 10-03-00827 and 10-03-90030, Russian Academy of Sciences (RAS) for grant in the framework of Complex Program DCHSM RAS "New metallic, ceramic, glass, polymeric, and composite materials", and Presidium of RAS (Program P21).

REFERENCES

[1] Pomogaylo A.D., Rosenberg A.S., Uflyand A.S. Metall Nanoparticals in Polymers. M.: *Khimiya,* 2000. 672 P.

[2] Gubin S.P., Koksharov Yu.A., Khomutov G.B., Yurkov G.Yu. Russ. *Chem. Revs.* 2005. V. 74 (6), P. 489–520.

[3] Gudoshnikov S., LiubimovB., Matveets L., Ranchinski M., Usov N., Gubin S., Yurkov G, Snigirev O., Volkov I. *Journal of Magnetism and Magnetic Materials.* 2003.V.258-259, P.54-56.

[4] Baker C., Ismat Shah S, Hasanain S.K.*Journal of Magnetism and Magnetic Materials.* 2004.V.280, №2-3, P.412-418.

[5] Kechrakos D., Trohidou K.N. *Applied Surface Science.* 2004. №226, P.261–264.

[6] Tackett R., Sudakar C. Naik R., Lawes G., Rablau C., VaishnavaP.P.*Journal of Magnetism and Magnetic Materials.* 2008. V.320, №21, P.2755-2759.

[7] Koichiro Hayashi, Wataru Sakamoto, Toshinobu Yogo. *Journal of Magnetism and Magnetic Materials.* 2009. V.321, №5, March P.450-457.

[8] Gufei Zhang, Potzger K., Shengqiang Zhou, Mücklich A, Yicong Ma, J.*Nuclear Instruments and Methods in Physics Research Section B: Beam Interactions with Materials and Atoms.* 2009. V.267, №8-9, 2009, P.1596-1599.

[9] Daiji Hasegawa, Haitao Yang, Tomoyuki Ogawa, Migaku Takahashi. *Journal of Magnetism and Magnetic Materials.* 2009. V.321, №7, 2009, P.746-749.

[10] Denkbas E.B., Kilicay E., Birlikseven C., Ozturk E. *Reactive and Functional Polymers.* 2002. V.50, P.225-232.

[11] Hyeon T. *Chem. Commun.* 2003. P. 927-934.

[12] Jun Y., Choi J., *Cheon J. Chem. Commun.* 2007. P.1203-1214.

[13] Kawamura M., Sato K. *Chem. Commun.* 2007. P.3404-3405.

[14] Muller J.-L., Klankermayer J., Leitner W. *Chem. Commun.* 2007. P.1939-1941.

[15] Dvorikova R. A., Antipov B.G., Klemenkova Z.S., Shanditsev V. A., Prokof'ev A.I., Petrovskii P.V., Rusanov A. L., Korshak Yu. V. *Polymer Science,* Ser.A 2005. Vol. 47, № 11, P.1135-1140.

[16] Dvorikova R. A., Nikitin L. N., Korshak Yu. V., Shanditsev V. A., Rusanov A. L., Abramchuk S.S., Khokhlov A. R.*Doklady Akademii Nauk* 2008. Vol. 422, №3, P. 334-338.

[17] Cooper A. I. *J. Mater. Chem.* 2000. Vol.10, P. 207-234.

[18] Nalawade S. P., Picchioni F., Janssen L.P.B.M. *Prog. Polym. Sci.* 2006. V. 31, P. 19–43.

[19] Reverchon E., Adami R. *J. of Supercritical Fluids. 2006.* V. 37, P. 1–22.

[20] Zhang Y., Erkey C. *J. of Supercritical Fluids.* 2006. V. 38, P. 252–267.

[21] Erkey C. *J. of Supercritical Fluids.* 2009. V. 47, P.517–522.

[22] Yang J., Hasell T., Wang W., Howdle S. M. *European Polymer Journal.* 2008. V. 44, P. 1331–1336.

[23] Said-Galiyev Ernest E., Vygodskii Yakov S., Nikitin Lev N., Vinokur Rostislav A., Gallyamov Marat O., Pototskaya Inna V., Kireev Vyacheslav V., Khokhlov Alexei R., Schaumburg Kjeld. *J. of Supercritical Fluids.* 2003. V. 27, P. 121-130.

[24] Nikitin Lev N., Gallyamov Marat O., Vinokur Rostislav A., Nikolaev Alexander Yu., Said-Galiyev Ernest E., Khokhlov Alexei R., Jespersen Henrik T., Schaumburg Kjeld. *J. of Supercritical Fluids.* 2003. V. 265, P.263-273.

[25] Yuvaraj H., Woo M. H., Park E. J., Jeong Y. T., Lim K. T. *European Polymer Journal.* 2008, V. 44, P. 637-644.

[26] Chen A.-Z., Kang Y.-Q., Pu X.-M., Yin G.-F., Li Y., Hu J.-Y. // Development of Fe_3O_4-poly(l-lactide) magnetic microparticles in supercritical CO_2. *Journal of Colloid and Interface Science.* 2009. V.330, P.317–322.

[27] Tsang S. C., Yu C. H., Gao X., Tam K. Y. // Preparation of nanomagnetic absorbent for partition coefficient measurement. *International Journal of Pharmaceutics.* 2006. V.327, P.139–144.

[28] Said-Galiyev E., Nikitin L., Vinokur R., Gallyamov M., Kurykin M., Petrova O., Lokshin B., Volkov I., Khokhlov A., Schaumburg K. // New chelate complexes of copper and iron: synthesis and impregnation into a polymer matrix from solution in supercritical carbon dioxide. *Industrial and Engineering Chem. Research.* 2000, V.39, P.4891-4896.

[29] Blackburn J. M., Long D. L., Cabanas A., Watkins J. J. // Deposition of Conformal Copper and Nickel Films from Supercritical Carbon Dioxide. *Science.* 2001. V.294, P.141-145.

[30] Vasilkov A., Naumkin A., Nikitin L. Volkov I., Podshibikhin V., Lisichkin G. //Ultrahigh molecular weight polyethylene modified with silver nanoparticles prepared by metal-vapour synthesis. *AIP Conference Proceedings.* 2008, V.1042, P.255-257.

[31] Nikitin L., Vasilkov A., Vopilov Yu., Buzin M., Abramchuk S., Bouznik V., Khokhlov A. // Making of metal-polymeric composites. *AIP Conference Proceedings.* 2008, V.1042, P.249-251.

[32] Nikitin L. N., Vasilkov A. Yu., Naumkin A. V., Khokhlov A. R., Bouznik V. M. //Metal-polymeric composites prepared by supercritical carbon dioxide treatment and metal-vapor synthesisin: *Success in Chemistry and Biochemistry: Mind's Flight in Time and Space, Volume 4* (A Festschrift in Honor of the 75th Birthday of Professor Gennady E. Zaikov), Editor: G.E.Zaikov, 2009, Nova Science Publishers, Inc. N.-Y., pp. 579-590.

[33] Dvorikova R. A., Nikitin L. N., Korshak Yu. V., Shanditsev V. A., Rusanov A. L., Abramchuk S.S., Khokhlov A.R. // New magnetic nanomaterials of hyperbranched ferrocenecontaining polyphenylenes prepared in liquid and supercritical carbon dioxidein: Quantitative Foundation of Chemical Reactions, Editors: Gennady E. Zaikov (N.M. Emanuel Institute of Biochemical Physics, Russian Academy of Sciences, Moscow, Russia) Tanislaw Grzegosz (Kaminski Institute of Natural Fibres, Poland) Lev N. Nikitin (A.N.Nesmeyanov Institute of Organoelement Compounds, Russia), 2009. *Nova Science Publishers,* Inc. N.-Y., P. 93-100.

[34] Sasaki Yu., Pittman Ch.U., Jr. // Acid-Catalyzed Reaction of Acetylferrocene with Trietyl Orthoformate. *J. Org. Chem.* 1973. V.38. № 21. P.3723-3726.

[35] Seyoum H.M., Bennet L.H. and Della Torre E. // Temporal and temperature variations of dc magnetic aftereffect measurements of Fe3O4 powders *J. Appl. Phys.* 2003. № 5, 2820-2822.

In: Chemical and Biochemical Kinetics: New Perspectives ISBN 978-1-61209-224-9
Editor: Gennady Zaikov

Chapter 14

The Structure of Polybutadienes and Butadiene-Acrylonitrile Copolymers

N. M. Livanova,* S. G. Karpova and G. E. Zaikov
Emanuel Institute of Biochemical Physics,
Russian Academy of Sciences, Moscow, Russia

Abstract

The paramagnetic probe method with the use of free radicals of different dimensions (2,2,6,6-tetramethyl-1-piperidinyloxy and 4-benzoate-2,2,6,6-tetramethyl-1-piperidinyloxy) has been employed to study the effect of the isomeric composition of butadiene units in polybutadienes and butadiene–acrylonitrile copolymers on the number and dimensions of ordered structures. The nature of density fluctuations and defective regions, that is, the regions in which the radicals are sorbed, has been ascertained. It has been shown that the ordered regions are composed of stereoregularchain fragments, while defective regions are enriched with butadiene isomers different from those present in prevailing amounts.

In [1], the micro- and supramolecular structure of ethylene–propylene–diene ternary copolymers was studied by IR and EPR spectroscopy. Paramagnetic probe measurements showed that the temperature dependences of correlation time τ_c plotted in Arrhenius coordinates for microblock copolymers with high degrees of isotacticity of propylene units demonstrate transitions relevant to the breakdown of supramolecular structures composed of not only ethylene sequences but also propylene sequences [1]. Later on, the existence of these transitions was confirmed by the thermomechanical analysis of the copolymers [2]. Under the assumption that the radical solubility and distribution over rotation frequencies are the same in amorphous regions of homo- and copolymers, the correlation time of a probe in ethylene–propylene–diene copolymers was calculated with a high accuracy on the basis of the additive contribution of each component corresponding to its weight fraction after deduction of the amounts of ordered structures. The latter were determined as the contents of fractions

* E-mail: livanova@sky.chph.ras.ru.

insoluble in toluene at room temperature. Thus, the paramagnetic probe method is a delicate instrument for investigating the molecular and supramolecular organization of elastomers.

Here, this method was applied for studying the molecular mobility in butadiene elastomers with different stereoisomerisms of units and in butadiene–acrylonitrile copolymers (acrylonitrile–butadiene rubbers (NBRs)). In most cases, they were mostly composed of isomeric butadiene units of one kind (*cis*-1, 4- in PBs or *trans*-1,4- in NBRs) (Table 1). The temperature dependence of the rotational mobility of free-radical probes with different dimensions (2,2,6,6-tetramethyl-1-piperidinyloxy (TEMPO) and 4-benzoate-2,2,6,6-tetramethyl-1-piperidinyloxy (BZONO)) in these elastomers was investigated [3, 4].

Table 1. Isomeric composition of butadiene units in PBs of different brands and butadiene–acrylonitrile copolymers

Elastomer	Isomer content*, %		
	trans-1,4	1,2	*cis*-1,4
cis -1,4-*PB*	0,89	0,6	98,5
SKD	3,8	2,5	93,7
SKDL	47	8,8	44,2
BNKS-18	82,0	8,2	9,8
BNKS -28	76,4	14,4	9,2
BNKS -40	93,0	4,4	2,6

* In the butadiene part of the polymer.

Table 2. Arrhenius parameters of probe rotation in polymers and τ_c values at 24.5°C

Elastomer	$\tau_c \times 10^{+10}$, s (24°C)	E_1	E_2	$\tau_o^{(1)}$, s	$\tau_o^{(2)}$, s	Temperature of the transition onset, K
		kJ/mol				
TEMPO						
cis-1.4-PB	0,89	33,5	56	$1,1\times10^{-16}$	$2,5\cdot10^{-20}$	323
SKD	0,5	24,4	54,1	$2,7\times10^{-15}$	$2,8\times10^{-20}$	314
SKDL	0,9	17,2	48,5	$8,5\times10^{-14}$	$4,4\times10^{-19}$	308
BNKS -18	7,4	33,8	54,6	$9,3\times10^{-16}$	$1,5\times10^{-19}$	311
BNKS -28	17,8	14,4	65,2	$5,6\times10^{-12}$	$1,2\times10^{-20}$	306
BNKS -40	23,4	20,1	76,4	$7,8\times10^{-13}$	$3,8\times10^{-22}$	317
BZONO						
cis-1.4-PB	4,0	44,3	33,0	$5,8\times10^{-18}$	$4,5\times10^{-16}$	313
SKD	5,0	44,0	35,8	$8,0\times10^{-18}$	$2,0\times10^{-16}$	309
SKDL	7,9	46,5	43,3	$5,0\times10^{-18}$	$1,7\times10^{-17}$	308

Note: E_i and $\tau_o^{(i)}$ are the effective activation energies and preexponential factors at temperatures below (i = 1) and above (i = 2) the temperature of the transition onset.

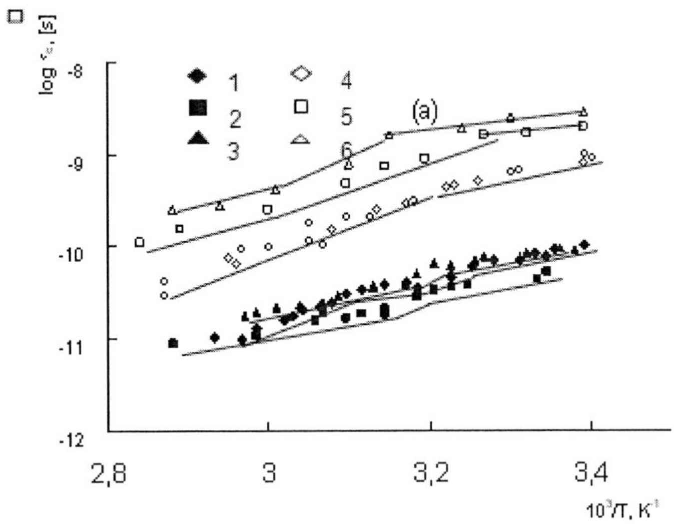

(a)

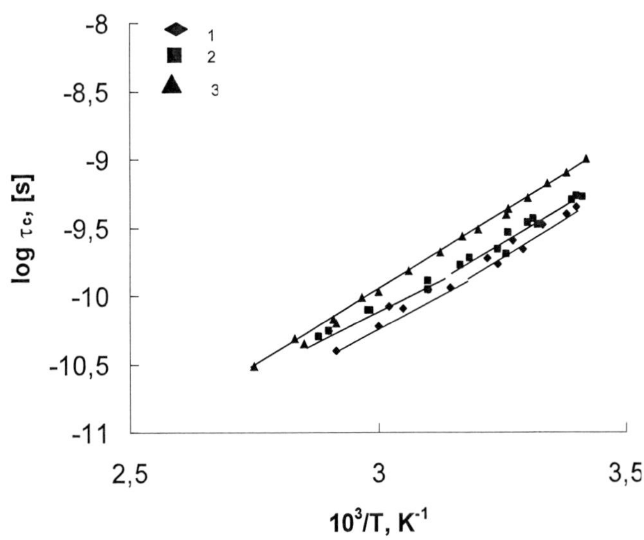

(b)

Figure 1. Temperature dependences of correlation times of the rotational mobility plotted for (a) TEMPO and (b) BZONO radicals in Arrhenius coordinates: (*1*) *cis*-PB, (*2*) SKD, (*3*) SKDL, (*4*) BNKS-18, (*5*) BNKS-28, and (*6*) BNKS-40.

Experimental

The objects of investigation were polybutadienes, such as *cis*-PB, SKD PB, and SKDL PB (a rubber synthesized in the presence of a lithium catalyst), with different stereoregularities of macromolecules and butadiene–acrylonitrile copolymers of the BNKS-18, BNKS-28, and BNKS-40 brands (acrylonitrile contents of 18, 28, and 40 wt %, respectively). The isomeric composition of butadiene units, that is, the contents of *trans*-1,4, *cis*-1,4, and 1,2-isomers, was determined by IR spectroscopy from bands at 967, 730, and 911 cm^{-1}, respectively, [5] through the use of film samples cast from solutions in CCl_4 or chloroform. The calculations were performed with the extinction ratios reported in [6]; the results are listed in Table 1.

EPR spectroscopy was used to study the rotational mobility, which is characterized by correlation time τ_c of TEMPO and BZONO radicals [3, 4], in a temperature range of 20–90°C with an EPR-V radiospectrometer. The elastomers were saturated with TEMPO from the gas phase at room temperature for three days. Then, samples were stored at room temperature for a few days. During storage, the radical was desorbed from large defective regions of the samples (pores, cavities, etc.), in which polymers have almost no effect on the rotational mobility of the radical. BZONO was incorporated into PB from solutions in acetone with a concentration of 10^{-3} mol/l with subsequent evaporation of the solvents at room temperature. The EPR spectra of the TEMPO probe in PBs and BNKS-18 show the classical triplet pattern. The spectra of both radicals measured in BNKS-28 and BNKS-40 at room temperature attest to a heterogeneous structure, that is, to the existence of fast and slow components. In the case of BZONO, the pattern of the spectra is typical of anisotropic rotation characterized by different frequencies of rotation about different molecular axes. The τ_c values determined at 24.5°C are listed in Table 2. The error in the measurements of τ_c is 5%.

RESULTS AND DISCUSSION

The effective activation energyE_{ef} was calculated for the rotation of the radicals (Table 2) from the temperature dependences of correlation time τ_c of TEMPO and BZONO in PBs and TEMPO in butadiene–acrylonitrile copolymers (Figure 1). The calculations were performed for different parts of the curves plotted in Arrhenius coordinates. These parts corresponded to different relaxation processes relevant to supramolecular structures. During approximation of the linear parts of the curves in Figure 1, correlation coefficients R^2 were 0.82–0.97.

For stereoregular polybutadienes (SKD rubbers) and NBRs, in the region of the high-elasticity plateau, relaxation spectrometry data indicate as many as three λ-relaxation processes relevant to the molecular mobility of structural elements that are larger than segments; that is, microblocks whose activation energy is equal to that of the viscous flow of the elastomers [7]. Hence, segments play the role of kinetic units of slow relaxation λ processes. The mechanism governing the molecular mobility of the microblocks consists in their formation and decomposition through the fluctuation attachment and detachment of segments as individual kinetic units [7]. Moreover, the thermomechanical analysis, the energy parameters of dissolution and diffusion of low-molecular-mass substances, and IR spectroscopy data suggest the existence of fluctuation structures in elastomers [8–12].

The presence of ordered microscopic regions and their size distribution is, as is shown below (and in [13]), governed by the microtacticity (the geometry similarity of chain fragments) and temperature. The molecular dynamics of polymer chain fragments located in a structural defect depends on the mobility of regular chain fragments comprising ordered structures. The transitions are explained by the unfreezing of the molecular mobility in fluctuation structures composed of units occurring in the prevailing configuration and by their rearrangements [7–13].

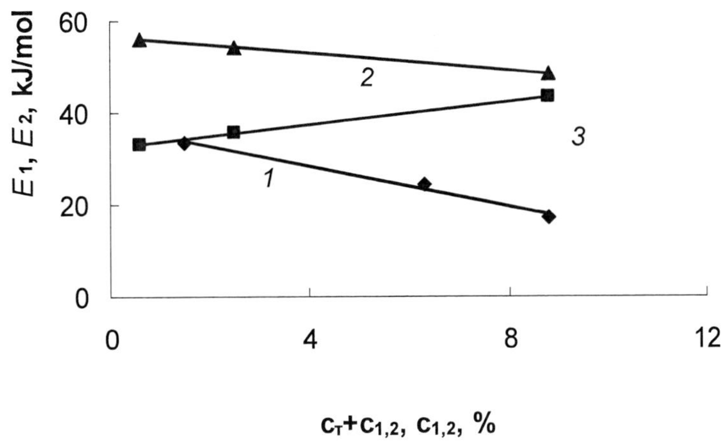

Figure 2. Effective activation energies (*1*) E_1 and (*2, 3*) E_2 of the rotational mobility of radicals as functions of the content of (*1*) *trans*-1,4- and 1,2-units ($c_T + c_{1,2}$) and (*2, 3*) 1,2-conformers ($c_{1,2}$): (*1, 2*) TEMPO and (*3*) BZONO.

Like any other low-molecular-mass substance, radicals are sorbed in the polymer regions with subsequent deteriorations of short-range order in the arrangement of chains [14]. Such regions most probably appear at the joints of different isomers of butadiene units, because there is no geometric similarity between them [13, 15]. When samples are stored for a few days at room temperature, TEMPO remains in the regions in which its desorption is hindered, that is, in regions with rather dense macromolecular packing but a free volume sufficient for accommodation of the radical. Defects in the chain stereoregularity may give rise to the appearance of such a free volume. The radical cannot penetrate regions with ordered structure and hindered molecular mobility.

Stereoregular PBs containing more than 99% *cis*-1,4-units form crystals with a maximum melting temperature of 12.5°C. When the content of *cis*-1,4-isomer is 98%, the melting temperature of the polymers is 0°C; at room temperature, their ordered structures are preserved [16]. At a *cis*-1,4-unit content of 90%, the maximum crystallinity of SKD is 55% (–37°C), while at 81%, the crystallinity is insignificant. It was shown that, during photochemical isomerization, crystallization occurs when the content of *trans*-1,4-units is higher than 75% [17]. However, when the content of *trans*-1,4-isomers is lower than 20%, *cis*-1,4-structures undergo low-temperature crystallization [18].

Experimental melting temperatures T_m depend on the contents of units occurring in different configurations and on the duration and temperature of crystallization and the rate of melting [16]. At a *cis*-1,4-unit content of 96.4%, the DTA method demonstrated the appearance of two overlapping endothermic peaks during rapid cooling and slow heating [19]. The DTA and dilatometry data depend on the frequency and randomness of the repetition of units with a structure different from the *cis*-1,4-configuration along a polymer chain [16]. Thus, the capability of ordering or crystallization is predetermined by the stereoregularity of elastomer chains. This conclusion is confirmed below by the EPR data.

For *cis*-PB and SKD and SKDL with different isomeric butadiene unit ratios, the effective activation energyE_1 of TEMPO radical rotation at temperatures below the transition linearly diminishes with an increase in the content of isomers present in small amounts, that is, *trans*-1,4- and 1,2-units (Figure 2). The total amount of *trans*-1,4- and 1,2- units in *cis*-PB and SKD is small (less than 10%) (Table 1); therefore, it is assumed that they are randomly distributed over polymer chains. According to [20], SKDL, which is synthesized in the presence of *n*-BuLi as a catalyst, contains blocks of cis and trans isomers connected by vinyl units. Hence, only *cis*-1,4-unit–1,2-unit–*trans*-1,4-unit joints are present in it. *cis*-1,4 and *trans*-1,4 isomers form fluctuation structures [11]. Thus, the regions with the lowest order are enriched with 1,2-units located between *cis*-1,4- and *trans*-1,4-isomers. Hence, the E_1 value of SKDL falls on the linear dependence of E_1 on the isomeric composition of butadiene units (Figure 2) if the content of 1,2-units (8.8%) (Table 1) is plotted on the abscissa axis.

The effective activation energy of probe rotation decreases in proportion to reduction in the stereoregularity of chains and the dimensions of ordered structures (Tables 1, 2; Figure 2). This fact agrees with the data reported in [16], where a PB sample synthesized in the presence of an $ALR_3/TiCl_4$-type catalyst was separated into fractions with different contents of *cis*-1,4-units by selective extraction.

In the case of *cis*-PB containing 98.5% *cis*-1,4-units and a content of *trans*-1,4- and 1,2-isomers of ~15%, the activation energy of the probe rotation in the initial region of the $\tau_c(T)$ curve (20–50°C) is 33.5 kJ/mol (Table 2). This result aggress with the data from [7], where the activation energy of the low-frequency relaxation processes, which are relevant to the mobility of microblocks of supramolecular structures and λ processes, for monodisperse PBs was found to be 34 kJ/mol.

In SKD containing 93.7% *cis*-1,4-units, structures composed of sequences of these units are found; however, because of a higher content of *trans*-1,4- and 1,2-units than that in *cis*-PB and, accordingly, a lower chain stereoregularity, the amount and dimensions of these structures are smaller than those in *cis*-PB. Therefore, the temperature of transition is lower (Table 2). The effective activation energy of TEMPO rotation in SKD (22.5 kJ/mol) is close to the activation energy of viscous flow determined for PB with a *cis*-1,4-unit content of 92% (23.5 kJ/mol) and the activation energy of sulfur diffusion in SKD (23.1 kJ/mol) [7, 10, 21].

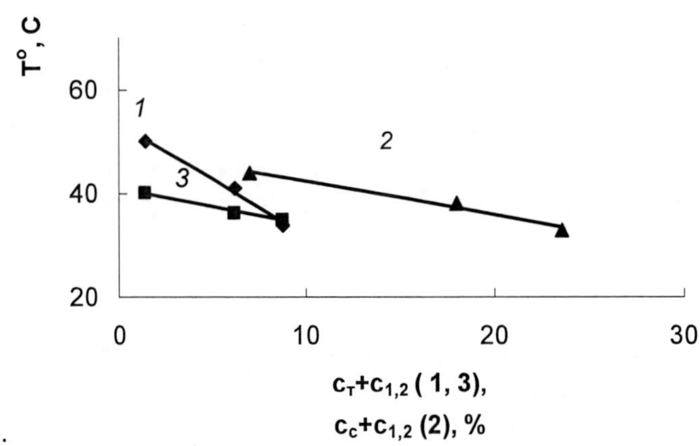

Figure 3. Transition temperatures of (*1*, *3*) PBs and (*2*) NBRs vs. the contents of (*1*, *3*) *trans*-1,4- and 1,2-units ($c_T + c_{1,2}$), (*2*) *cis*-1,4- and 1,2-butadiene units ($c_c + c_{1,2}$): (*1*, *2*) TEMPO and (*3*) BZONO.

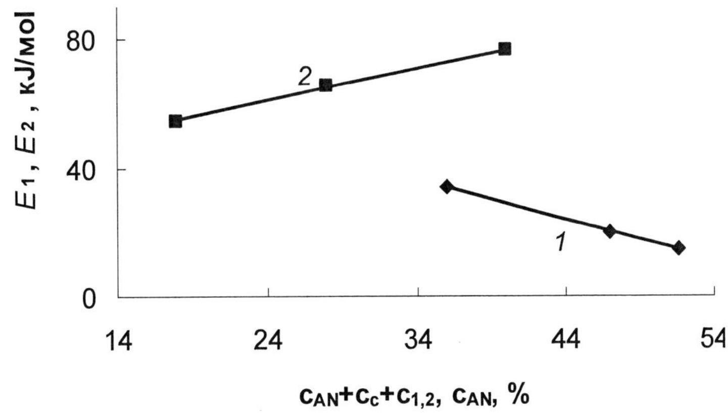

Figure 4. Effective activation energies (*1*) E_1 and (*2*) E_2 of radical TEMPO rotation vs. total contents of (*1*) acrylonitrile + *cis*-1,4- + 1,2-units ($c_{AN} + c_c + c_{1,2}$), (*2*) c_{AN}.

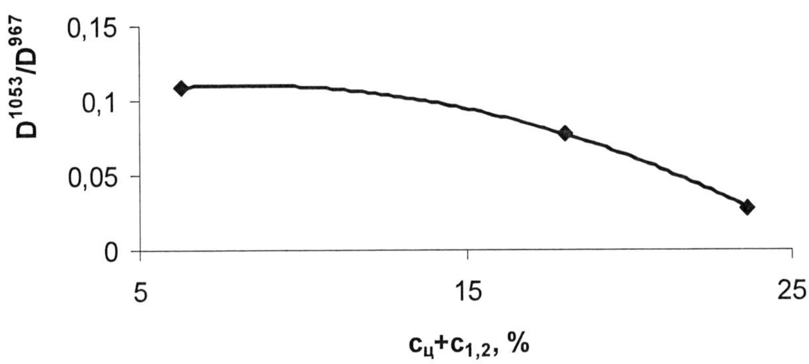

Figure 5. Optical density ratio for bands at 1053 and 967 cm^{-1} characterizing the fraction of ordered structures in NBRs vs. total content of *cis*-1,4- and 1,2-units ($c_c + c_{1,2}$).

Thus, the E_{ef} value of SKD is equal to the activation energy of the viscous flow of PB with a close isomeric composition. In SKD, the radical is sorbed by regions with defective chain microtacticity that are composed of the same sequences of isomers as those in *cis*-1,4-PB.

In SKDL containing almost equal amounts of *cis*-1,4- and *trans*-1,4-isomers, the effective activation energy of radical rotation is 17.5 kJ/mol at temperatures below the transition temperature (35°C). The obtained transition temperature agrees with [11].

Hence, if the regular parts of chains are composed of ordered structures, polymer chains located in unordered regions (the regions of radical sorption) are enriched with butadiene isomers that are present in smaller amounts.

Similarly to E_1, the transition temperature linearly decreases with an increase in the content of *trans*-1,4- and 1,2-isomers in PBs (Figure 3). This fact lends support to the conclusion that, at low temperatures, the temperature dependence of the radical mobility is governed by the dimensions of stereoregular fragments of chains and related ordered structures.

The effective activation energy of TEMPO radical rotation, E_2, corrresponding to the region above the transition temperature is higher than E_1 and slightly declines according to a linear relationship with an increase in the 1,2-isomer content (Figure 2). Hence, the transition is associated with the breakdown of primary structures and separation of chains related to an increase in segmental mobility; as a result, only vinyl units containing side double bonds influence the radical rotation.

In butadiene–acrylonitrile copolymers with different contents of polar units and isomer ratios of the butadiene comonomer (Tables 1, 2), the effective activation energies of TEMPO probe rotation that correspond to the initial part of Arrhenius dependence$\tau_c(T)$ likewise linearly decrease with an increase in the amount of all defects in the regularity of the butadiene component (∑AN-units + *cis*-1,4-units + 1,2-units) (Figure 4). Energy E_2 corresponding to the region above the transition temperature linearly increases with the content of acrylonitrile units in the copolymers, that is, with the concentration of polar units exhibiting a high cohesion energy, because of a strong dipole–dipole interaction of nitrile groups, which noticeably increase chain rigidity. Thus, the transition improves the segmental mobility so much that chains are separated and the effect of butadiene unit isomers becomes unnoticeable against the background of the strong interaction between the polar groups, which strongly reduce chain flexibility.

Above the transition temperature, the mobility of TEMPO in PBs and NBRs is influenced by side vinyl or nitrile groups. In both cases, the larger the amount of defects in the stereostructure of the butadiene part of a chain, the lower the effective energy of radical rotation.

Figure 5 shows the dependence of the intensity ratio between the band at 1053 cm^{-1} attributed to the *trans*-1,4-units contained in crystallites and the band at 967 cm^{-1} that characterizes the total content of *trans*-1,4-isomers [5] on the total content of 1,2- and *cis*-1,4 isomers.As their content increases, the order of *trans*-1,4-units diminishes. BNKS-28 and BNKS-40 have the least and the most ordered structures, respectively. As was shown in [16], at a *trans*-1,4-unit content above 75%, the crystallinity is observed at room temperature. The total amorphization attests to a random distribution of *trans*-1,4-units. Hence, at a high content of acrylonitrile units, BNKS-40 contains the greatestamounts of structures formed by both strong dipole–dipole bonds between nitrile groups and *trans*-1,4-butadiene sequences [7, 12, 22].

The BZONO probe, which is characterized by a high molecular mass, an asymmetric shape, and an anisotropic rotation, was incorporated into the elastomers from solutions; therefore, the sample structure may be different from the initial one. TEMPO molecules occupy defects that exist in a polymer. As regards BZONO, it seems to partly forms defects in the course of solvent evaporation. This fact may explain different patterns of the Arrhenius $\tau_c(T)$ dependences plotted for TEMPO and BZONO radicals (Figure 1).

When BZONO is used, relaxation transitions are distinctly observed for *cis*-PB and SKD (Figure 1, Table 2). Linear $\tau_c(T)$ dependences or small differences between E_{ef} values in the first and second regions suggest the absence of ordered structures or their insignificant effect on the radical mobility (their absence or small dimensions in SKDL).

In contrast to TEMPO, the BZONO radical occurs in the environment of long disordered defective butadiene fragments of chains. This circumstance seems to influence the E_{ef} value.

Equal E_1 values, which are observed for BZONO in PBs (Figure 2) irrespective of the stereoregularity of elastomer chains, may be interpreted as a consequence of the probe location in large defects possessing the same structure, while close temperatures of the relaxation transitions (Figure 3), which are lower than those for TEMPO, may be considered to be due to a smaller size of ordered structures. The E_2 value of BZONO slightly increases with an increase in the content of 1,2-isomers of butadiene units (similarly to TEMPO) (Figure 2). This finding may be explained by certain hindrances to the rotational mobility of the large probe from the side units of chains.

The above results lead us to draw the following conclusions. A small TEMPO radical provides more complete information on the structure of polymers and the nature of defective regions in them. The incorporation of a probe from solutions is accompanied by changes in the initial structural organization of elastomers. For some systems, this phenomenon may be of interest. Radicals are sorbed by regions of a polymer matrix that contain butadiene units. Defective regions comprise chain fragments with alternating butadiene units of different isomeric structures. Their mobility is influenced by regular chain fragments composing ordered structures, a fact that additionally confirms the existence of the latter.

REFERENCES

[1] N. M. Livanova, S. G. Karpova, and A. A. Popov, *Polymer Science,* Ser. A 45, 238 - 243 (2003) [Vysokomol. Soedin., Ser. A 45, 417-423 (2003)].

[2] N. M. Livanova, Yu. V. Evreinov, A. A. Popov, and V. A. Shershnev, *Polymer Science,* Ser.A 45, 530-536 (2003) [Vysokomol. Soedin., Ser. A 45, 903-912 (2003)].

[3] A. M. Vasserman and A. L. Kovarskii, *Spin Labels and Probes in Physical Chemistry of Polymers* (Nauka, Moscow, 1986) [in Russian].

[4] A. M. Vasserman, A. L. Buchachenko, A. L. Kovarskii, and M. B. Neiman, *Vysokomol. Soedin.,* Ser. A 10, 1930–1936 (1968).

[5] J. Dechant, R. Danz, W. Kimmer, and R. Schmolke, *Ultrarotspektroskopische Untersuchungen an Polymeren* (Akademie, Berlin, 1972; Khimiya, Moscow, 1976).

[6] N. V. Kozlova, F. F. Sukhov, and V. P. Bazov, *Zavod. Lab.* 31, 968-970 (1965).

[7] G. M. Bartenev, *Structure and Relaxation Properties of Elastomers* (Khimiya, Moscow, 1979) [in Russian].

[8] V. A. Kargin, Z. Ya. Berestneva, and V. G. Kalashnikova, *Usp. Khim.* 36, 203-216 (1967).

[9] V. P. Lebedev, *Usp. Khim.* 47, 127-151(1978).

[10] G. M. Bartenew, *Plast. Kautsch.* 17, 727-730 (1970).

[11] L. V. Sokolova, O. A. Chesnokova, O. A. Nikolaeva, and V. A. Shershnev, *Vysokomol. Soedin.,* Ser. A 27, 352-361 (1985).

[12] V. S. Grishin, I. A. Tutorskii, and I. S. Yurovskaya, *Vysokomol. Soedin.,* Ser. A 20, 1967-1973 (1978).

[13] V. N. Kuleznev, Polymer Science, Ser. B 35, 1156 (1993) [*Vysokomol. Soedin.*, Ser. B 35, 1391-1402 (1993)].

[14] Yu. A. Shlyapnikov, S. G. Kiryushkin, and A. P. Mar'in, *Antioxidation Stabilization of Polymers* (Khimiya, Moscow, 1986) [in Russian].

[15] V. G. Kulichikhin, G. V. Vinogradov, A. Ya. Malkin, et al., *Vysokomol. Soedin.,* Ser. B 10, 739-742 (1968).

[16] *The Stereo Rubbers*, Ed. by W. M. Saltman (Wiley, New York, 1979), Vol. 2.

[17] M. Sh. Yagfarov, *Vysokomol. Soedin.,* Ser. A 10, 1264-1269 (1968).

[18] E. A. Collins and L. A. Chandler, *Rubber Chem. Technol.* 39, 193 (1966).
[19] M. Golub, *J. Polym. Sci.* 25, 373-377 (1957).
[20] V. D. Mochel, *J. Polym. Sci.,* Part A-1 10, 1009-1018 (1972).
[21] V. S. Grishin, I. A. Tutorskii, and E. E. Potapov, *Vysokomol. Soedin.,* Ser. A 16, 130-135 (1974).
[22] L. V. Sokolova, O. A. Konovalova, and V. A. Shershnev, *Kolloidn. Zh.* 44, 716-721 (1982).

In: Chemical and Biochemical Kinetics: New Perspectives ISBN 978-1-61209-224-9
Editor: Gennady Zaikov

Chapter 15

Relation between the Cell-free DNA Content and the Lipid Peroxidation in the Blood Plasma of Mice under Damaging Action

Lyudmila N. Shishkina,* Mikhail A. Klimovich, Mikhail V. Kozlov, and Margarita A. Smotryaeva
Department of the chemical and biological process kinetics,
Emanuel Institute of Biochemical Physics of Russian Academy of Sciences,
Moscow, Russia

Abstract

The detail analysis of interrelation between the cell-free DNA content and the LPO intensity in blood plasma of intact different species mice and under damaging factors (low-toxicity chemical agent, industrial toxicant and ionizing radiation at low doses) were done. It is shown that both the LPO intensity and cell-free DNA content have the high sensitivity to the action of chemical factors and ionizing radiation at the low doses. However, the scale of its changes is substantial dependence on the factor origin and also its concentrations and the dose rateradiation.

Keywords: mice, blood plasma, cell-free DNA, lipid peroxidation, X-rays, chemicalagent, black liquor, low doses

1. Introduction

As known, the most acute response to various damaging factors is produced by the hematopoietic system, which participates in maintaining the homeostasis on the level of

* E-mail: shishkina@sky.chph.ras.ru.

organism. Changes in the hematopoietic system under the action of the chemical factors and ionizing radiation are of prime importance in forming both early and long-term biological consequences after the damaging actions. Among the most sensitive parameters to the radiation action especially at the low doses under the low dose rate there are parameters of the physicochemical regulatory system of the lipid peroxidation (LPO) which is maintained the steady-state of the LPO in the tissues of the intact laboratory animals [1]. At present the cell-free DNA content is widely suggested as the marker in the case of many diseases including pathologies in pregnancy [2-8]. A function and origin of the low-molecular cell-free DNA are extremely studied under different diseases and conditions, but a fact is turned one's attention thatthere are simultaneously changes of the LPO regulatory system parameters. Since the LPO initiators (hydroxyl radical, superoxide anion radical, peroxides) cause damages to not only lipids but also DNA bases [9], one might expect that there is a relationship between damages to the genetic structures and/or DNA biosynthesis and the LPO intensity. This assumption is in accordance with the data about the decreased oxidative DNA damage in human white blood cells after the antioxidant supplementation in diet [10, 11]. Moreover, in our investigations there were the reverse correlation between the rate constant for alkaline elution of the peripheral blood lymphocyte DNA and the oxidation products in blood plasma for the reference animals and irradiated mice at the dose of 15 cGy under the different dose rate and the direct correlation between the cell-free DNA content and the oxidation products in blood plasma of mice SHK whose drinking water was daily supplement with the industrial chemical toxicant at various concentrations [12]. However the linear regression coefficients of these correlations were distinguishable between the reference and experimental groups of mice.

The aim of this work is the detail analysis of interrelation between the cell-free DNA content and the LPO intensity in blood plasma of intact mice and under damaging factors.

2. Materials and Methods

Experiments on Animals

The X-irradiation at low doses and also the low-toxicity chemical agents and the industrial toxicant were used as damaging factors. The 130 white outbreed mice, 88 mice SHK and 45 Balb/c mice made a choice as the experimental animals because they characterize the various antioxidant status of tissues and different sensitivity to ionizing radiation [13, 14]. To modify the antioxidant status experiments were performed at the different seasons. Animals during experiments were in the special cages per 10 mice, a food and fresh drinking water mice received daily.

As the low-toxicity chemical agents we used polyoxyethylensorbitanmonooleate (Tween 80) which is usually applied for the hydrophobic agent administration and the relative low-toxicity chemical solvent acetone. The experiments were carried out on Balb/c mice (males, weight 16 – 20 g before the experiments, n = 13) during October and November. The mice were divided into two groups. The first group (8 mice) was used as the intact biological control. The Tween 80 (Ferak Berlin, FRG) at the dose of 30 mg/kg as 0.3% solution in 10% mixture of acetone in water was administration intraperitoneally at the second group of mice

(5 mice). The solution of 0.3% Tween 80 in the 10% aqueous solution of acetone (special purity grade) was prepared immediately before the experiment.

As the industrial toxicant we used the black liquor (a residue in cellulose pulping process), which were daily added in drinking water. The experiments were carried out on 60 mice SHK (females) during January and February. Mice were divided into 4 groups per 15 mice. Mice in experimental groups were received a drinking water which the black liquor was supplement at concentrations of the 0.005 (group I); 0.015 (group II) and 0.05% (group III), correspondingly. Mice of group IV were the intact age biological control and received a common drinking water without supplement.

The single X-irradiation of white outbreed mice (females) at the doses less than 0.1 mGy under the decreased dose rate during exposure was performed by a Microwave electric-Electron Cyclotron Resonance Accelerator which was developed and in detail described in Ref. [15, 16]. The absence of the microwave component during exposure was specially controlled. The experiments were carried out on 130 white outbreed mice (females) during November and December (experience 1) and September and October (experience 2). Mice per 10 in group were irradiated in special container where the each mouse was singly and could easily drive. The exposure time was about 18 minutes.

Decapitation of mice was performed within 1 week after X-rays (30 mice, experience 1) and 1 month after damaging actions all another experiments. Decapitation of mice from the same animal group which served as the intact age biological control was simultaneously performed with the experimental groups. After decapitation of mice, blood was collected in test tubes treated by 5% solution of sodium citrate. The erythrocytes from the blood plasma were separated by centrifugation. All parameters of intact mice were determined for 1 or 2 animals. Mice in experimental groups were divided per 1 – 3 animals.

Biochemical Analyses

The content of LPO products, which have reacted with 2-thiobarbituric acid (TBA-reactive substances, TBA-RS) was determined spectro phometrically at the wavelength 532 nm with Beckman Du-50 (Austria) or KFK-3 (Russia) instruments by the method described in [17] with adding 10 μl of the 0.01% 4-methyl-2,6–di*tert.*butylphenol (BHT) solution in ethanol. Protein was determined according to [18].

The content of cell-free DNA concentration in blood plasma was determined by fluorescence technique [19, 20] with the help of a Hitachi M 850 (Japan) spectrofluorimeter. Blood plasma (1 ml) was mixed with 1 ml of 4',6-diamidino-2-phenylindole dihydrochloride (DAPI) solution (0.4 μg/ml). The mixture was incubated for 30 min at 37° C since the whole binding of DNA and DAPI was within 15 – 30 min according to [21]. The fluorescence measurements were carried out on a spectrofluorimeter in a 0.02 M solution of disodium EDTA, pH 10. The wavelength of the exiting UV-radiation was 338 nm, the fluorescence spectra were measured at a wavelength of 457 nm, slit was 5 nm, the scanning rate was 60 nm/min. The amount of DNA (μg/ml) in the sample was determined by calibration curve, which was done by the native DNA of murine liver.

The data were processed by a commonly used variational statistic method and by the KINS program given in [22]. The variability of indices was evaluated as ratio of mean square error of average mean to average mean for group expressed as a percentage. The significance of differences were evaluated by Student *t*- criterion.

Table 1. The TBA-reactive substances amount and DNA content in the blood plasma of the white outbreed mice (females) at the different season

Parameter	September	October (beginning)	October (the end of month)	November	December
[TBA-RS], nmol/mg of protein	0.076 ± 0.007 (n* = 4)	0.0765±0.004 (n = 2)	0.172± 0.005^{x} (n = 2)	0.148±0.014^{x} (n = 6)	0.082 ± 0.013 (n = 10)
[DNA], μg/mg of protein	0.037± 0.003 (n = 10)	0.0415 ± 0.0075 (n = 4)		0.060±0.009^{xxx} (n = 6)	0.0615±0.008^{xx} (n = 10)

* - Number of replicate measurements. Significant differences from the September data (^{x}p < 0.01; ^{xx} p < 0.02; ^{xxx} p < 0.03).

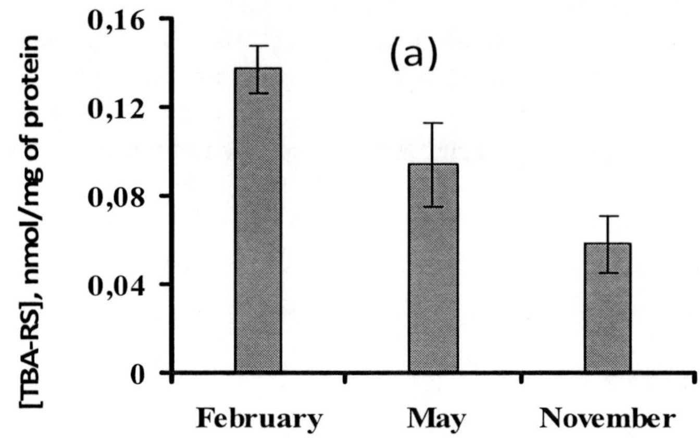

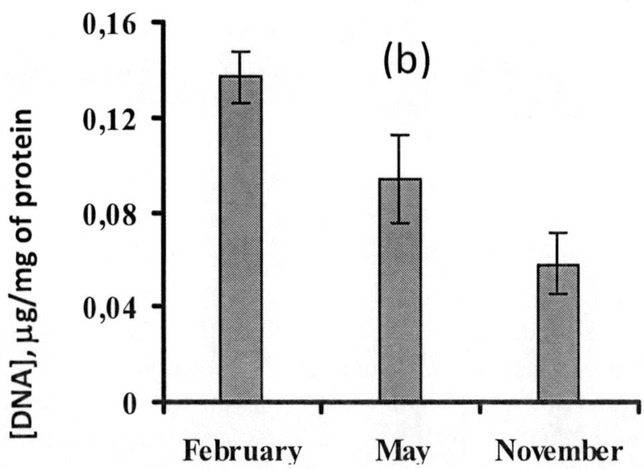

Figure 1. The TBA-reactive substances content (a) and the cell-free DNA amount (b) in the blood plasma of mice Balb/c (males) at the different season. Significant differences from the control group: * - p < 0.1, ** - p < 0.01.

3. Results and Discussion

As shown, the LPO is the common physiological process which is need for the functioning all types of biological membranes, cells and tissues [23, 24]. Earlier it was obtained that there are the substantial season differences both in the antioxidant activity of the spleen and liver lipids of mice SHK and Balb/c [13, 14] and the liver lipid composition and also the LPO products in the lipids and liver homogenate of mice Balb/c [25]. It is well known that LPO intensity is evaluated by the TBA-reactive substances content in a complex biological system [26]. Indeed, the existence of the reverse correlation between the TBA-reactive substances content in the liver homogenate and thelipid antioxidant activity of mice SHK liver which is presented in [27] is a accordance with this conclusion. As the first step of our investigation, a possibility of the season changes of the TBA-RS content and cell-free DNA amount in the blood plasma of mice was studied. The data obtained are presented in Figure 1 for mice Balb/c and in Table 1 for the white outbreed mice. It is seen that studied parameters are for certain differed in dependence on season in which experiments were done.

Thus, the TBA-reactive substances content in the blood plasma of mice Balb/c which are characterized a more low antioxidant status of tissues than the white outbreed mice [13, 14]may differ from 10.9 to 18.3 times in autumn and spring as compared with winter season. Besides the DNA amount in the blood plasma of these groups of mice is also substantially higher under performing experiments in winter. The lowest levels both the LPO intensity and cell-free DNA content in the blood plasma of white outbreed mice reveal in September (Table 1). Earlier the high variability of the TBA-RS content was also found in the blood plasma of mice SHK (males) [28]. It is interesting to compare the average values of the cell-free DNA content and TBA-reactive substances amount in the blood plasma for groups of mice. This comparison is presented in Figure 2 for mice Balb/c (males). It is seen that this interrelation has the complex character. Moreover, the scale of interrelation between the cell-free DNA content and TBA-reactive substances amount in the blood plasma of the different species of mice is also unequal and substantially depends not only on the LPO intensity in the murine blood plasma. That assumption allows us to advance data which are presented in Figure 3. As could be seen, the increase of the LPO intensity in blood plasma doesn’t always lead to the rise of the interconnection between the cell-free DNA and TBA-reactive substances amounts.

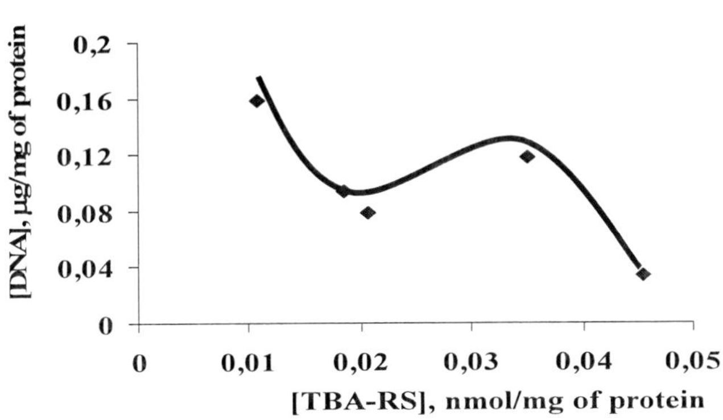

Figure 2. The interrelation between the average values of the cell-free DNA content and TBA-reactive substances amount in the blood plasma for groups of intact mice Balb/c.

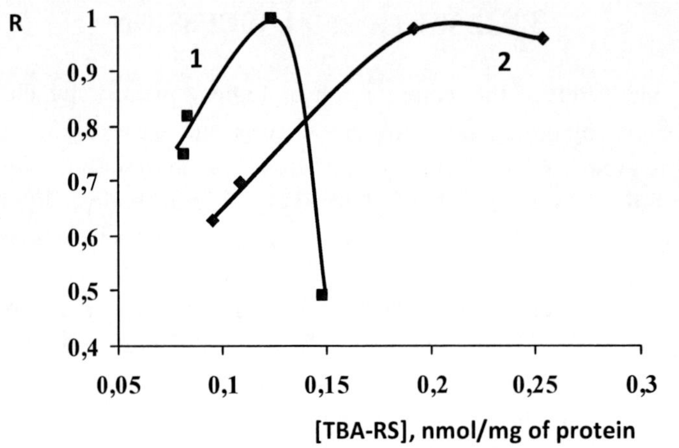

Figure 3. Dependence of the correlation coefficient for the interrelation between the cell-free DNA content and LPO product amount on the LPO intensity in the blood plasma of the white outbreed mice (females) (1) and mice SHK (females) (2).

Table 2. The TBA-reactive substances amount and DNA content in the blood plasma of mice within 1 month after the damaging action

Species of mice	Variant of experiment	[TBA-RS], nmol/mg of protein	[DNA], μg/mg of protein
Mice Balb/c (males)	The intact age control (n* = 4)	0.021 ± 0.006	0.078 ± 0.016
	Administration of 0.3% Tween 80 in 10% water acetone (n = 2)	0.046 ± 0.012	0.0325 ± 0.0055[xx]
Mice SHK (females)	The intact age control	0.092 ± 0.007 (n = 5)	0.102 ± 0.008 (n = 9)
	The black liquor in drinking water at the concentration of0.005%	0.146 ± 0.029 (n = 5)	0.144± 0.012 (n = 3)[x]
	0.015 %	0.143 ± 0.029 (n = 5)	0.116 ± 0.016 (n = 5)
	0.05 %	0.075 ± 0.007 (n = 5)	0.096 ± 0.025 (n = 5)
Mice SHK (females) [27]	The intact age control (n = 6)	0.1915 ± 0.028	0.072 ± 0.008
	X-rays at the dose of 16 (n = 3)	0.42 ± 0.20	0.117 ± 0.043

* Number of replicate measurements. [x] – Significant differences from the control mice ($p < 0.02$). [xx] - Significant differences from the control mice ($p < 0.06$).

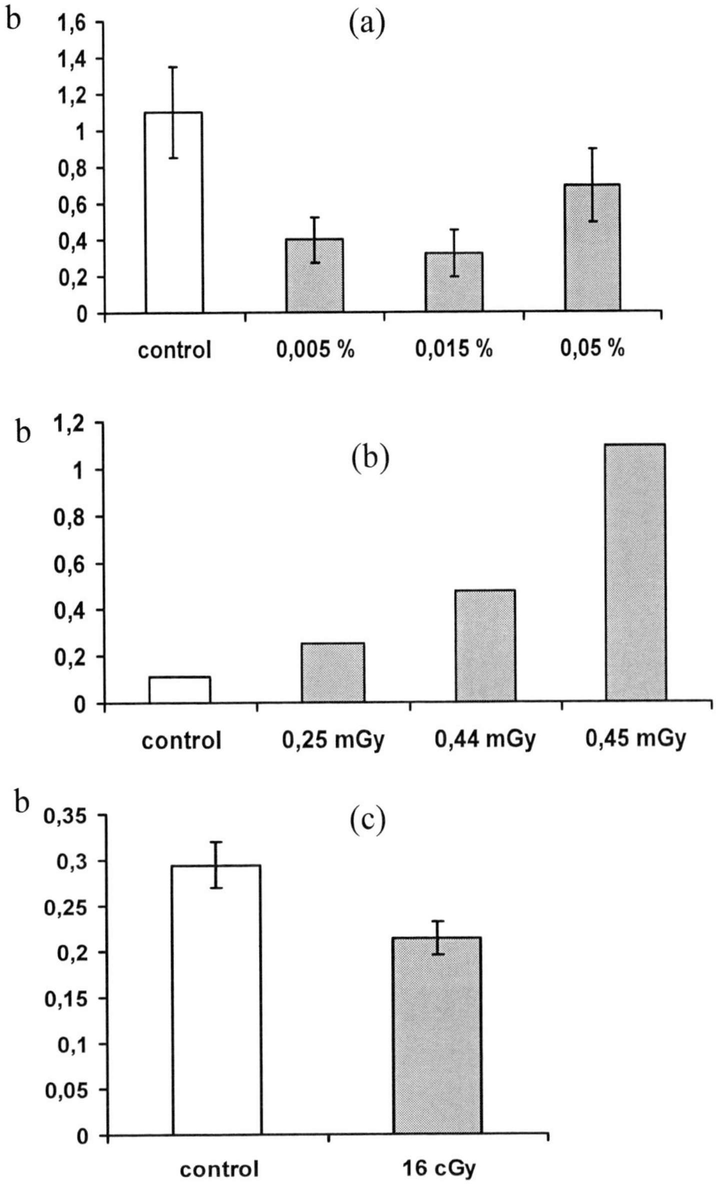

Figure 5. Effect of damaging factors on values of the linear regression coefficients of the direct correlations between the cell-free DNA content and TBA-reactive substances amount in the blood plasma of mice with 1 month after action. (a)– presence of the black liquor in the drinking water, (b) – X-rays irradiation at the low doses under changing the dose rate, (c)– X-rays irradiation at the dose of 16 cGy.

Since the damaging factors cause disturbances in the LPO regulatory system functioning [1, 13, 14, 27] it might be expected that there are revealed changes both the level of the cell-free DNA content and the LPO intensity and the scale and direction of the relation between these indices in the blood plasma of mice after damaging actions. The obtained data which are presented in Table 2 and also in Figure 4 are confirmed this assumption.

As seen, the administration of the low-toxicity chemical agents results to the reliable diminution of the cell-free DNA content, however the LPO intensity reveal the tendency to the rise (Table 2). The presence of the black liquor in the drinking water of mice causes the same dynamic changes of both studied parameters, but the effect depends on the toxicant concentration. The most effect is obtained by the black liquor concentration of 0.005 % (Table 2). There is a tendency to the increase of both investigated indices after the X-ray irradiation of mice at the low dose because of their high heterogeneity in the blood plasma of the experimental group of mice (Table 2). It is need to note that high heterogeneity of the system response is one of the distinctive peculiarities under a weak actions [12, 29]. Moreover, the response of a complex biological system to weak influence can be substantial dependence on the initial values of its parameters. Data which are presented in Figure 4 are a accordance with this conclusion. Besides, the scale of the TBA-reactive substances content is dependent not only on its initial value in the blood plasma of mice in the control group, but the dose rate change during exposure (Figure 4 a and b).

CONCLUSION

In this paper the season changes both the LPO product content and cell-free DNA amount in the blood plasma of the different species of mice are revealed. It is found that both the LPO intensity and cell-free DNA content have the high sensitivity to the action of chemical factors and ionizing radiation at the low doses. However, the scale of its changes is substantial dependence on the factor origin and also on its concentrations and the radiation intensity. This can be seen in Figure 5 where are presented the effect of the different damaging factors on values of the linear regression coefficients for the direct correlations between the cell-free DNA content and TBA-reactive substances amount in the blood plasma of mice. Obviously, the reliable season changes of the cell-free DNA content and the LPO intensity in the murine blood plasma, its high and unequal sensitivity to the damaging factors and substantial dependence on the initial values in the control mice and also the intensity and/or concentration of factor would be expected to be also a reason for the existence of several hypotheses about the cell-free DNA origin.

ACKNOWLEDGMENTS

This work was supported by the International Science and Technology Center Program, Projects No 547-98 and 1032, and also The Program of Fundamental Research of Presidium of the Russian Academy of Sciences “Fundamental sciences – for Medicine” (2006 – 2008).

REFERENCES

[1] L.N. Shishkina, E.V. Kushnireva, M.A. Smotryaeva, “A new Approach to the Assessment of Biological Consequences of Exposure to Low-dose Radiation”, *Radiation Biology. Radioecology*, vol. 44, no. 3, pp. 289-295, 2004 (in Russian).

[2] R. Gonzales, J.M. Siva, G. Domingues, et al., "Microsatellite alternations and TP53 Mutations in Plasma DNA of Small Cell Lung Cancer Patients: Follow-up study and Prognostic Significance", *Annual Oncology*, vol. 11, pp. 1097-1104, 2000.

[3] V.G. Vladimirov, I.N. Vasil'eva, L.A. Sharova, "Extracellular DNA and Its Significance for Current Medicine", *Clinical Medicine and Physiopathology*, no. 1-2, pp. 110-119, 1998 (in Russia).

[4] B. Burwinkel, M.W. Kilimann, "Unequal Homologous Recombination Between Line-1 Elements as a Mutational Mechanism in Human Genetic Disease", *Journal of Molecular Biology*, vol. 277, pp. 513-517, 1998.

[5] Y.M.D. Lo, M.S. Tein, T.K. Lau, et al., "Quantitative analysis of fetal DNA in maternal plasma and serum: implications for noninvasive prenatal diagnosis", *American Journal ofHuman Genetics*", vol. 62, pp. 768-775, 1998.

[6] L.Y. Chan, T.N. Leung, K.C. Chang, et al., "Serial analysis of fetal DNA concentrations in maternal plasma in late pregnancy", *Clinical Chemistry*, vol. 49, pp. 678 – 680, 2003.

[7] W.D. Bianchi, "Circulating fetal DNA: its origin and diagnostic potential – a review", *Placenta*, vol. 18, pp. 93-101, 2004.

[8] S. Grill, C. Rusterholz, R. Zanetti-Dallenbach, et al., "Potential markers of preeclampsia – a review", *Reproductive Biology and Endocrinology*, 7:70doi:10.1186/1477-7827-7-70, 2009.

[9] A.R. Proteggente, T.G. England, C.A. Rise-Evans, B. Halliwell, "Iron Supplementation and Oxidative Damage to DNA in Healthy Individuals with High Plasma Ascorbate", *Biochemical and Biophysical Research Communications*, vol. 288. pp. 245-251, 2002.

[10] M. Porrini, P. Riso, "Lymphocyte Lycopene Concentration and DNA Protection from Oxidative Damage is Increased in Women after a Shot Period of Tomato Consumption". *Journal of Nutrition*, vol. 130, pp. 189-192, 2000.

[11] P. Moller, S. Loft, "Oxidative DNA damage in human white blood cells in dietary antioxidants intervention studies", *American Journal of Clinical Nutrition*, vol. 76, pp. 303-310, 2002

[12] L.N. Shishkina, M.A. Smotryaeva. "Relationship of Membrane and DNA Damage with Lipid Peroxidation under Weak Influence", *Biophysics*, vol. 45, no. 5, pp. 818 – 826, 2000.

[13] E.B. Burlalova, A.V. Alesenko, E.M. Molochkina, et al., *Bioantioxidants in Radiation Damage and Tumor Growth*. Moscow, Nauka, 1975 (in Russian).

[14] L.N. Shishkina, E.B. Burlakova, "The value of antioxidant properties of lipids in radiation damage and membrane repair", In: *Chemical and Biological Kinetics. New horizons* / Eds. E.B. Burlakova, S.D. Varfolomeev, Leden, Boston: VPS, pp. 334 – 364, 2005.

[15] K.F. Sergeichev, D.M. Karfidov, N.A. Lukina, "Electron Cyclotron Resonance Acceleration of Electrons to Relativistic Energies by a Microwave Field in a Mirror Trap", *Plasma Physics Reports,* vol. 33, no. 6, pp. 455-473, 2007.

[16] D.M. Karfidov, K.F. Sergeichev, "Characteristics of X-rays from ECR-discharge in a Mirror Trap", *Applied Physics*, no. 6, pp. 102-105, 2007 (in Russian).

[17] T. Asakawa, S. Matsushita, "Coloring Conditions of Thiobarbituric Acid Test for Determination Lipid Hydroperoxides", *Lipids*, vol. 15, no. 3, pp. 137–140, 1980.

[18] R. Itzhaki, D.M. Gill, “A micro-biuretic method for estimating proteins”, Analytical Biochemistry, vol. 9, no. pp. 401 – 410, 1964.

[19] J. Kapuscinski, B. Skoczylas, “Simple and rapid fluorimetric method for DNA microassay”, *Analytical Biochemistry*, vol. 83, pp. 252-257, 1977.

[20] J. Kapuscinski, B. Skoczylas, “Fluorescent complexes of DNA with 4’,6-diamidino-2-phenyl indole.2HCl or DCl 4’,6-dicarboxyamide-2-phenyl indole”, *Nucleic Acids Research*, vol. 5, no. 10, pp. 3775-3799, 1978.

[21] N.V. Blukhterova, M.A. Smotryaeva, K.E. Kruglyakova, “Comparative Study of the Effect of 1-Methyl-1-Nitrosourea and 1,3-Dimethyl-1-Nitrosourea on DNA of Tumor Cells *in vitro* and *in vivo* by the Method of Alkali Elution”, *Biological Bulletin*, no. 3, pp. 276-281, 1996 (in Russian).

[22] E.F. Brin, S.O. Travin., “Modeling of chemical reaction mechanism”, *Chemical Physics Reports*, vol. 10, no. 6, pp. 830-837, 1991 (in Russian).

[23] E.B. Burlakova, N.G. Khrapova, “Membrane lipid peroxidation and the natural antioxidants”, *Achievements of Chemistry*, vol. 54, no. 9, pp. 1540-1658, 1985 (in Russian).

[24] K. Hensley, K.A. Robinson, P. Gabbina, et al., “Reactive Oxygen Species, Cell Signaling and Cell Injury”, *Free Radical Biology and Medicine*, vol. 28,no. 10, pp. 1456-1462, 2000.

[25] M.V. Kozlov, V.V. Urnysheva, L.N. Shishkina, “Interconnection of Parameters of Regulation System of Lipid Peroxidation and Morphophysiological Parameters of Mouse Liver”, *Journal of Evolutionary and Biochemical Physiology*, vol. 44, no. 4, pp. 470-475, 2008.

[26] V.E. Kagan, O.N. Orlov, L.I. Prilipko, “*Problems of Analysis of Lipid Peroxidatoion Products*”. In: Itogi nauki I tekhniki. Moscow: VINITI Akad. Nauk SSSR. Ser. Biofizika, 1986 (in Russian).

[27] L.N. Shishkina, Ye.V. Kushnireva, M.A. Smotryaeva, “The combined effect of surfactant and acute irradiation at low dose on the lipid peroxidation process in tissues and DNA content in blood plasma of mice”, *Oxidation Communications*, vol. 24, no. 2, pp. 276– 286, 2001.

[28] L.N. Shishkina, Ye.V. Kushnireva, O.F. Bespal'ko, N.V. Polyakova, :The Role of Antioxidative Status in Development of the Consequences of Biological Action Under Low Doses and Low Intensity Irradiation”, *Radiation Biology. Radioecology*, vol. 40, no. 2. pp. 162-7 (in Russian).

[29] E.B. Burlakova, “Effect of ultra low radiation doses”, *Herald of Russian Academy of Sciences*, vol. 64, pp. 425-431, 1994.

In: Chemical and Biochemical Kinetics: New Perspectives
Editor: Gennady Zaikov
ISBN 978-1-61209-224-9

Chapter 16

INHIBITION MECHANISM OF NATURAL AND SYNTHETIC ANTIOXIDANTS IN THE PRESENCE OF PHOSPHOLIPIDS AND HYDROPEROXIDES

L. I. Mazaletskaya*[*], *N. I. Sheludchenko and L. N. Shishkina
Emanuel Institute of Biochemical Physics of Russian Academy of Sciences, Moscow, Russia

ABSTRACT

Resultsof investigations of the phospholipids (PL) and hydroperoxide influence on the antioxidant action efficiency of the synthetic phenolic antioxidants (InH) –2,6-di-*tert*-butyl-4-methylphenol (BHT), 4-methoxyphenol (MOP), 4-*tert*-butylphenol; natural InH - α-tocopherol (TP), quercetin (Q) and dihydroquercetin (QH_2) and hydrated quinoline – 2,2,4-trimethyl-6-ethoxy-1,2-dihydroquinoline (ethoxyquin, EQ), which are performed by the model of the initiated oxidation of methyl oleate are generalized. It is established that the PL additives lead to both a increase of the oxidation rate within the induction period (W_τ) and a alteration of the induction period duration (τ).According to the increase in W_τ for mixtures with PL studied InH is graded by the following sequence: $QH_2 > Q > EQ > TP > MOP > BHT$, that is conformed by the differential UV spectroscopy data. The value of τ increases in the presence of PL for TP, MOP and EQ, it is unchangeable for BHT and substantially decreases for QH_2 and Q. Reasons leading to a alteration of the InH efficiency in the PL presence are discussed.

Keywords: inhibition, synthetic antioxidant, α-tocopherol, quercetin, dihydroquercetin, phospholipids

AIM AND BACKGROUND

Biological activity of compounds is mostly due to their ability to take part in the regulation of oxidative processes in organism. The intensity of oxidative processes is particularly reduced, if the system contains a special group of substances – antioxidants

(InH), the reactivity of which depends on the conditions of oxidation, the nature of oxidation substrates, physicochemical properties of the InH [1-6] and is distinguished from the systems of varying degrees of complexity.

In this work we consider the influence of natural components of cells - phospholipids (PL) and the primary molecular products of oxidation - hydroperoxides on the inhibiting efficiency of a number of InH by the initiated oxidation of methyl oleate.

EXPERIMENTAL

Following InH were used: synthetic phenols – 2,6-di-*tert*-4-methyl-butylphenol (BHT), 4-methoxyphenol (MOP), 4-tert-butylphenol (TBP), natural InH – α-tocopherol (TP), quercetin (Q), dihydroquercetin (QH_2) and hydrated quinoline – 2,2,4-trimethyl-6-ethoxy-1,2-dihydroquinoline (ethoxyquin, EQ), a molecule of which contains an active aminogroup. Studied PL were phosphatidylcholine from egg yolk (PC, «Sigma»), dipalmitoyl-phosphatidylcholine (DPPC, «Sigma») and lecithin («Serva»). According to the TLC, the lecithin is a mixture of natural lipids, containing 56.6% PL, among which 96.6% was PC.Methyl oleate oxidation (333 K), initiated by azobisizobutironitrile (AIBN), was controlled by the kinetic oxygen uptake. Methyl oleate in the mixture with inert dissolvent chlorobenzene (1:1) and the dissolved initiator were termostated. Then InH and PL were added. The initial rates of oxidation, as well as the duration of the induction period (τ),were determined from the kinetic curves of oxygen uptake by method described in Ref 7. The induction period in the presence of InH was equal interval from the beginning of the experience until projection the point of intersection of two straight lines for which tg $\alpha_1 = 2$ tg α_2. The first line is corresponded to a extension of a straight of the oxygen uptake, when the reaction rate is constant after the complete consumption of InH. The second line is tangent to the kinetic curve of the oxygen uptake in the point, in which the reaction rate is twice less than that in the absence of inhibitor.

RESULTS AND DISCUSSION

The simplest kinetic scheme, describing the initiated oxidation process in the presence of InH includes following reactions:

(i) $AIBN \rightarrow 2\ r\cdot(RH) \rightarrow 2R^{\cdot}$

(1) $R^{\cdot} + O_2 \rightarrow RO_2^{\cdot}$

(2) $RO_2^{\cdot} + RH \rightarrow ROOH + R^{\cdot}$

(3) $ROOH \rightarrow RO^{\cdot} + OH^{\cdot}$

(6) $RO_2^{\cdot} + RO_2^{\cdot} \rightarrow ROOR + O_2$

(7) $InH + RO_2^{\cdot} \rightarrow In^{\cdot} + ROOH$

(8) $In^{\cdot} + RO_2^{\cdot} \rightarrow$ molecular products

(9) $In^{\cdot} + In^{\cdot} \rightarrow$ molecular products

(10) $In^{\cdot} + RH + O_2 \rightarrow RO_2^{\cdot} + InH$

(10′) $In^{\cdot} + ROOH \rightarrow RO_2^{\cdot} + InH$

(11) InH + ROOH→molecular products
(12) InH + O_2→ $In^{\cdot}$+$HO_2^{\cdot}$

As seen, along with the reactions which lead to break of the oxidation chain (7, 8), InH and their radicals may be involved in side reactions of the propagation chain (10, 10, 12). Reactions 11 and 12 lead to the "unproductive" consumption of InH and in such a way the InH concentration, involved in the process of the oxidation inhibition, is decreased. The efficiency of InH depends on a complex of reactions with their participation, the contribution of which changes a dependence on the oxidation conditions. The mathematical analysis of kinetic curves of oxygen uptake in the presence of the InH additives, which was conducted by the software package KINS [8] testifies about the contribution of side reactions in the mechanism of the InH inhibition. Calculations show that kinetic curves of oxygen uptake in the presence of BHT or Q (methyl oleate, 333 K) correspond to the mechanism, when there is the interaction of these InH with peroxyl radicals mainly [9]. Hence, the role of side reactions in the inhibition mechanism for these InH is very small. In the contrary, the character of kinetic curves in presence TP, MOP and EQ suggests the InH involvement in the side reactions.

Experimental data, which evidences about the interaction of TP and EQ with hydroperoxide (ROOH) and the lack of interaction between Q and ROOH, are consistent with the results of the mathematical analysis. The reaction between Q and ROOH was studied by means of the consumption of Q under anaerobic conditions at 333 K. It is established, that the consumption rate of Q (W_Q) at $[ROOH]_0$ = constis independent on $[Q]_0$ in the range of Q concentrations, provided the linear chain termination. However, W_Q increases with the growth in the $[ROOH]_0$ concentration at the fixed initial concentration of Q. The results obtained indicate that Q does not practically interact with ROOH under experimental conditions. The consumption of Q is due to its interaction with free radicals forming during the thermal decay of ROOH.

The ability of TP to decompose methyl oleate ROOH was confirmed, when TP at the concentration of 4.8×10^{-2} mol/l was added to oxidized methyl oleate at room temperature: the concentration of ROOH decreased with 4.5×10^{-2} to 3.7×10^{-2} mol/l [10]. The result of the interaction of TP with ROOH, apparently, is reduction of the duration of the induction period with increasing initial concentration of ROOH. So, under $[ROOH]_0 = 2 \times 10^{-4}$ mol/l τ_{TP} = 61 min, while under$[ROOH]_0 = 1.2 \times 10^{-2}$ mol/l τ_{TP} = 48 min. Thus, the presence in the system of the primary oxidation product – ROOH, may reduce the InH effectiveness even under conditions of moderate temperature (333 K) and high rate of generation of free radicals, that is due to the reaction between InH and ROOH. Individual PL almost had no influence on the reaction rate of initiated oxidation methyl oleate: kinetic curves in the presence and absence of PL coincided in all cases (Figure 1, the dotted line). However, adding PL to InH alter the effectiveness of their action. Comparison of the kinetic curves of oxygen uptake during the oxidation of methyl oleate in the presence of TP (Figure 1, curve *1*) and the mixture of TPand lecithin (Figure 1, curve *2*) shows that the addition of lecithin alters the kinetic curves of oxygen uptake, leading to an increased rate of oxidation during the induction period (Wτ). At the same time the duration of the induction period is increased. The different effect of the PL additives on the values Wτ and τ can lead to indefinitive conclusions under the evaluation of the inhibiting action effectiveness for such mixtures. So, if the inhibition effectiveness is evaluated by oxygen uptake within the induction period or Wτ, then the action of the mixture can be described as antagonistic. When the comparative analysis is performed by the induction period duration, mixture can be regarded as synergistic. Point of intersection of curves 1 and 2 (Figure 1) formally corresponds to the additive action of the components in the mixture and isolates the antagonistic field from the synergistic action. Two other stadies PL – PC and DPPC exert a similar influence on the efficiency action of TP. The increase of Wτ in the presence of PL was also observed for BHT, MOP, TBP, Q, QH_2, EQ. The most powerful influence of PL on the Wτ change was detected for the flavonoids.

To estimate changes in the efficiency of InH within the induction period the ratio of the efficiencies of inhibition in the presence of InH and its mixture with PL was calculated: (W_0/W_{InH}): $(W_0/W_{mix}) = W_{mix}/W_{InH}$, where W_0 is the rate of initiated oxidation, W_{InH} and $W_{mix}$are the rate of oxidation into the induction period in the presence of InH and the mixtures of InH with the PL, respectively. The higher ratio W_{mix}/W_{InH} is due to the greater reduction in the effectiveness of the InH in the presence of the PL within the induction period. As seen from Table 1, where the corresponding ratio for the mixtures of different InH and PL is presented the ratio magnitude depends on the InH structure and the PL concentration. On the basis of these data antioxidants may be arranged accoding to their efficiency in presence of PL as the following series

$QH_2 > Q > EQ > TP > MOP > BHT$.

Besides, value of W_{mix}/W_{InH} depends on the initial concentration of the components in the mixture and increases with increasing concentrations of InH and PL (see Table 2, Figure 3).

Dependence of W_{mix}/W_{InH} on $[PL]_0$ at a fixed concentration of InH, is a straight line, which cut off the segment on the ordinate axis, equal to 1. This is illustrated by data obtained during the oxidation of methyl oleate in the presence of mixtures of PC and EQ (Figure 3).

Proceed from the ability of several antioxidants interact with the hydroperoxides, it can be assumed that the degree of the substrate oxidation may also change the effectiveness of a mixture of InH with the PL. Effect of initial concentration of hydroperoxide ($[ROOH]_0$) inmethyl oleate was examined by the example of mixtures of TP and MOP (Table 3).

Table 1. The values of W_{mix}/W_{InH} for antioxidants with different structures

Antioxidant	[InH]0 ×104, mol/l	[PL]0, mg/ml	Wmix/WInH
Dihydroquercetin (QH2)	2	1.25*	10.2
Quercetin (Q)	2	1.25*	2.4
Ethoxyquin (EQ)	2.7	1.25**	1.1
Ethoxyquin	2.7	2.5**	3.0
α-Tocopherol (TP)	2.7	2.5**	2.1
4-Methoxyphenol (MOP)	2.7	2.5**	1.8
BHT	2.7	2.5**	1.4

* Lethitin.

** PC.

Indeed, the increase in $[ROOH]_0$ leads to an increase in the W_{mix}/W_{InH} (Table 3), while the dependence of W_{mix}/W_{InH} on $[ROOH]_0$ is a straight line, the segment of which $W_{mix}/W_{InH} > 1$ at $[ROOH]_0 = 0$(Figure 4). The magnitude of this ratio indicates how many times increases W_{mix} compared with W_{InH}.

Change of Wτ in the PL presence may be due to the interaction of components in the mixture. The reaction of the interaction of PL with InH was studied by means of the spectrophotometrical method at room temperature. The electron absorption spectra of lecithin with TP are presented in Figure 5. The changes, observed in the differential absorption spectra of mixtures, indicate on the interaction between components in the mixture. Similar results were obtained for other PL – PC in mixture with monophenols [11]. It was found that the scale of the changes of the absorption spectra of mixtures is determined by the degree of the steric hindrance for OH-group in molecules of monophenols, which are arranged in series TP > MOP > BHT. This series is coincided with the sequence of changes for W_{mix}/W_{InH} values, obtained for phenols containing one OH-group during methyl oleate oxidation. In Ref 12 it is also established, that the interaction of Q with PL leads to the formation of complexes and accompanied by a change in the electronic and the conformational structure of the Q molecule.

Table 2. The values of W_{mix}/W_{InH} in the oxidation of methyl oleate (333 K) at different initial concentrations of the components in the mixture

Antioxidant	[InH]0 ×104, mol/l	[PC]0, mg/ml	Wmix/WInH
Ethoxyquin	0.9	2.5	1.4
Ethoxyquin	2.7	2.5	3.0
Ethoxyquin	2.7	1.0	1.5
Ethoxyquin	2.7	1.25	1.1
α-Tocopherol	0.9	2.5	1.2
α-Tocopherol	2.7	2.5	2.4
α-Tocopherol	2.7	1.25	1.2

Methyl oleate: chlorobenzene = 1:1, $W_i = 1 \times 10^{-7}$ mol/l s.

Table 3. The values of W_{mix}/W_{InH} in the oxidation of methyl oleate (333 K)

Antioxidant	[ROOH]0×102, mol/l	Wmix/WInH
α-Tocopherol	0.02	1.2
α-Tocopherol	0.13	1.3
α-Tocopherol	0.61	1.8
α-Tocopherol	1.2	2.3
4- Methoxyphenol	0.13	1.3
4- Methoxyphenol	1.1	1.8

Methyl oleate: chlorobenzene = 1:1, Wi = 1 × 10-7 mol/l s, [InH]0 = 2.7 × 10-4 mol/l, [PC]0 = 2.5 mg/ml.

Table 4. The values of the relative residual chemiluminescence $(I/I_0)_0$ during initiated oxidation of ethylbenzene in the presence of 4-*tert*-butylphenol and its mixtures with PC

Temperature, K	Wi×107, mol/l s	[TBP]0 ×104 mol/l	[PC]0, mg/ml	(I/I0)0
333	0.5	2.9	–	0.266
333	0.5	2.9	3.1	0.436
353	1.0	0.63	–	0.17
353	1.0	0.63	0.06	0.19
353	1.0	063	0.15	0.256
353	1.0	0.63	0.28	0.308
353	1.0	0.63	0.34	0.339
353	1.0	0.63	0.4	0.31
353	1.0	0.63	1.25	0.448
353	1.0	0.63	1.7	0.526
353	1.0	0.63	2.5	0.545
353	1.0	0.63	4.25	0.717

Table 5. The duration of the induction period (τ) during the oxidation of methyl oleate (333 K), inhibited by the mixtures of quercetin and lecithin, $[Q]_0 = 2 \times 10^{-4}$ mol/l, methyl oleate: chlorobenzene = 1:1, $W_i = 1 \times 10^{-7}$ mol/l s

$[Lecithin]_0$, mg/ml	–	0.1	0.3	0.63	1.25
τ, min	67	58	41	40	38

Reduction of the inhibiting actioneffectiveness of InH in the presence of PL allow us to suggest, that the interaction between them results to the formation of complexes, which are less active or inactive in the inhibition and are in equilibrium with the initial components of the mixture. In this case the concentration of InH, that is not bound into the complex, is determined by the equation:

$$[InH] = [InH]_0/(1 + Kp\,[PL]_0), \quad (1)$$

where *K*p is equilibrium constant of the complex formation. The assumption, that the formation of the inactive in the inhibition complex leads to a decrease in the effective concentration of InH, can satisfactorily describe the experimentally revealed increase the initial rate of oxidation for the mixtures of PL and antioxidants. For the calculation we used data obtained during oxidation of model hydrocarbon - ethylbenzene in the presence of TBP and its mixtures with the PC, by means of a chemiluminescence technique (Table 4).

The choice of TBP was determined by the moderate value of the rate constant of its interaction with peroxyl radicals, equal $k_7 = 2.9 \times 10^4$ l/mol s (333 K) [13]. Besides, the registration of the chemiluminescence intensity for mixtures of InH and PL is easily carried out by means of the chemiluminescence for a itself InH. As found, I/I_0 (where I_0 andI are the intensities of chemiluminescence without additives and in the presence InH and PL) as well as W_{mix}/W_{InH}, gains with the increase of the PL concentration in the mixture (Table 4).

The changing values of the I/I_0 was calculated taking into account the formation of complex in dependence on $[PL]_0$. The rate of the inhibited oxidation under a mixed-chain breakage condition, in which is realized for mixtures of TBP with a PC and included the reactions $RO_2^{\bullet}$ + InH and $RO_2^{\bullet} + RO_2^{\bullet}$, is described by the equation

$$W_0/W_{mix} - W_{mix}/W_0 = fk_7\,[InH]_0/(k_6W_i)^{0,5}, \quad (2)$$

where W_0 is the rate of initiated oxidation of ethylbenzene in the absence of additives InH, k_6 is the reaction rate constant of square termination. As shown in Ref. 14, $W_0/W = (I_0/I)^{0.5}$. In the calculation it was assumed that [PC] = const, because in the reaction mixture [PC] >> [InH]. The final eguation obtained by taking into account (I) has the form:

$$[(I_0/I)^{0.5} - (I/I_0)^{0/5}]^{-1} = (k_6Wi)^{0.5}\,(1 + Kp[PL]_0)/fk_7[InH]_0 \quad (3)$$

As can be seen from Figure 6, the experimentally obtained values I/I_0 (Table 4) is linearized satisfactorily in coordinates of equation (III). Thus, the assumption about the formation of the inactive in inhibition complex allows us to describe satisfactorily the experimentally observed regularities, corresponding to the initial stages of reaction. Analysis of the chemiluminescence curves showed that by the PC additive to the antioxidants there is the some reduction for the tangent of the maximal slope (β) along with increase of the chemiluminescence intensity. So, for 2.9×10^{-4} mol/l TBP in presence of 3.1 mg/ml PC, β is decreased from $1.84 \times 10^{-4}\ s^{-1}$ to $1.4 \times 10^{-4}\ s^{-1}$. Similar results were obtained for MOP. Reduction of β, along with an increase of I/I_0 indicates that the complex formation, apparently, leads to a decrease in antiradical activity of InH in the mixture with PL. This conclusion is consistent with the data of Ref. 15-17, where the observed decrease in antiradical activity (k_7) of InH in the reaction with peroxyl radicals of tertiary amines as compared to k_7 for the same InH with peroxyl radicals of hydrocarbons is found. This result is associated with the formation of hydrogen bond between the H-atom of the hydroxyl group of InH and nitrogen-containing oxidation substrate.

Thus, the decrease of inhibiting efficiency of InH by addition of the PL at the initial stages of oxidation can be caused by reduction of the effective concentration of InH due to the complex formation and/or a decrease in antiradical activity of the product of the component interaction.

As already noted, in addition to increasing W_τ PL additive to InH can change the duration of τ. According to the influence of PL on the duration of τ for the different InH, the latter can be divided into three groups.

The first group includes EQ, TP and MOP, for which the induction period in the presence of PL is increased. At the same time scale of the effect depends on the nature of the PL: for a mixture of TP + PC the value $(\tau_{mix} - \tau_{InH})/\tau_{InH} = 4\%$, whereas for a mixture of TP and lecithin, this value is 15% Influence of the PL on τ was established for another class of natural antioxidants – polihydroxynaphtoquinons [18]. It could be assumed that the increase in τ for this group of InH takes place due to the fact thatformation of complex leads to a decrease not only the rate of inhibition reactions, but also the rate of side reactions with the participation of InH. Another reason for the prolongation of inhibitory action of mixtures compared to itself,

InH may be the regeneration of the InH mediated by the amino group of the tertiary amines. This conclusion is based on data [15, 17], according to which the retardation of tertiary amines oxidation leads to an increase of the stoichiometric inhibition coefficient, which is caused by repeated break of the oxidation chains.

Representatives of the second group are hindered phenols, for which the rate of side reactions is negligible. So, the PL additive to BHT has almost no impact on the value of τ.

For the third group, which includes flavonoids– Q and QH_2, is observed a sharp decrease in the induction period. So, the value of τ for Q, calculated from the kinetic curves (Figure 2, curve *1*), is $\tau = 67$ minutes, while the mixture Q with lecithin (1.25 mg/ml) leads to reduction τ to 37 min (Figure 2, curve *3*). QH_2 is very effective (Figure 2, curve *2*) in the absence of lecithin ($\tau = 82$ min), but its mixture with lecithin almost completely loses the activity and only slightly retards the oxidation process (Figure 2, curve *4*). For this reason, the determination of the induction period from the kinetic curve of oxygen uptake in the presence of a mixture of QH_2 with lecithin is not possible. For mixtures of Q with PL change of τ depends on the amount of added PL (lecithin).The most dramatic decrease in τ is observed within the concentration range of 0.1-0.3 mg/ml, a further increase in the amount of lecithin in a mixture practically does not change τ (Table 5). Thus, flavonoids differ from monophenols by the results of the PL influence. For the mixture of flavonoids and PL there is not only the most significant increase in the rate of oxidation within the induction period, but a substantial decrease in τ, i.e. a mixture of flavonoids with PL can be classified as antagonistic at all criteria.

The differences, observed for the effecteveness of mixtures of flavonoids and monophenols with PL, may be related to the fact that flavonoids, in contrast to monophenols, are more hydrophilic, and may be localized whithin the polar region of the lecithin bilayer (or micelles) [19] under their interaction with PL, while the TP is localized into the hydrophobic region. Since the PL form direct micelles in organic solvents [20], the flavonoids are located inside micelles or bilayer, formed by the PL molecules.Perhaps, the hindered access of radicals to InH is due to the flavonoid location inside micelles or bilayer formed of the PL molecules that leads to the reduction of the inhibition interval and to the increase of the initial oxidation rate, as well as the decrease of the antiradical activity. It should be noted that among the studied flavonoids, QH_2 is the more hydrophilic than Q, and, apparently, the reduction of the QH_2 effectiveness in the PL presence is more considerable for this reason. This assumption is consistent with the results of Ref. 20,where is revealed, that Q had ahigherefficiency compared with the more hydrophilic QH_2under their effects on the free radical induced hemolysis of erythrocytes and the platelet aggregation because of their different ability to interact with the cell membranes.

CONCLUSIONS

The effectiveness of the inhibiting action of antioxidants, depending on their involvement in side reactions may decrease with increasing content of hydroperoxides during the initiated oxidation at moderate temperatures. The PL additives to antioxidants lead to an increase in the rate of oxidation within the induction period. The scale of this effect increases with increasing initial concentration of the components in the mixture and the hydroperoxides content in the oxidation substrate. According to the PL influence on the duration of the induction period InH can be divided into three groups:

- antioxidants, for which there are an increase of the induction period, - α-tocopherol, 4-metoxyphenol, etoxyquin,
- antioxidants, for which the induction period is not changed - BHT,
- antioxidants, for whichthe induction period is decreased - quercetin and dihydroquercetin.

Dependence of the InH efficiency of on the nature of phenol and the PL is necessary to take account of their use in biologically active additive content.

References

[1] N.M. Emanuel, J.N. *Laskowskaya: Inhibition of oxidation of fats.* Pishchepromizdat, Moscow, 1961, 355 p.

[2] N.M. Emanuel, G.E. Zaikov, Z.K. Maizus: Oxidation of Organic Compounds. *Effect of Medium.* Percamon Press, Oxford, 1984, 650 p.

[3] V.A. Roginsky: Phenolic antioxidants. The efficiency and reactivity. *Nauka,* Moscow, 248 (1988).

[4] E.B. Burlakova, S.A. Karshakov, N.G. Khrapova: *Khim. Physics,* 14 (10), 151 (1995).

[5] E.T. Denisov, V.A.Azatian: Inhibition of chain reactions. *Chernogolovka,* 268 (1997).

[6] E.T.Denisov, T.G. Denisova: Handbook of Antioxidants; Bond Dissociation Energy; *Rate Constants; Activation Energy and Enthalpies of Reactions* (2th ed.): Boca Raton: London, New York, Washington: CRC Press, 2000, p. 290.

[7] N.M. Emanuel, G.P. Gladyshev,E.T. Denisov, V.F. Tsepalov, V.V. Kharitonov, K.B. Piotrowskii: *The order of testing chemical compounds as stabilizers of polymeric materials.* Preprint, Chernogolovka, 1976. p. 11.

[8] E.F. Brin, S.O.: *Travin Khim. Physics,* 10 (6), 830 (1991).

[9] N.V. Khrustova, L.N. *Shishkina: Kinetics and Catalysis.*45 (6), 848 (2004).

[10] L.I. Mazaletskaya, N.I. Sheludchenko, L.N. Shishkina: *Neftekhimiya,* 48 (2), 105 (2008).

[11] E.B. Burlakova, L.I. Mazaletskaya, N.I. Sheludchenko, L.N. Shishkina: *Izv. Sciences. Ser. Khim.,* (6), 1053 (1995).

[12] R.R. Sharafutdinova, J.G. Afanasiev, S.N. Zagidullin, R.S. Nasibullin: *Sat abstracts and reports in 11 All-Conference,* Nalchik, 28 June-2 July 2004 Yoshkar-Ola: Izd MarSTU. p.296. (2004).

[13] G.V. Karpukhina, Z.K. Mayzus, L.I. Matienko: *Neftekhimiya,* 6 (4), 603 (1966).

[14] V.J. Shlyapintoh, O.N. Karpukhin, L.M. Postnikov, I.V. Zakharov, A.A. Vichutinsky, V.F. Tsepalov: Chemiluminescence methods for the study of slow chemical reactions. *Nauka,* Moscow, 1966. p. 300.

[15] G.A. Kovtun, A.L. Aleksandrov: *Izv. USSR. Ser. Khim.* (1), 38 (1977).

[16] U.Ya. Samatov, A.L. Alexandrov, I.R. Ahunov: *Izv. USSR. Ser. Khim.,* 2254 (1978).

[17] F.N. Latypova, A.L. Alexandrov, S.S. Zlotsky, D.L. Rakhmankulov: *Izv. USSR. Ser. Khim.,* 951 (1979).

[18] L.V. Boguslavskaya, E.B. Burlakova, E.A. Koltsova, O.B. Maksimov, N.G. Khrapova: *Biophysics,* 35 (6), 928 (1990).

[19] P.Pedrielli, G. F.Pedulli, L. H. Skibsted: Antioxidant Mechanism of Flavonoids. Solvent Effect on Rate Constant for Chain-Breaking Reaction of Quercetin and Epicatechin in Autoxidation of Methyl Linoleate. *J. Agric. Food Chem.,* 49, 3034 (2001).

[20] L. Ross C. Barclay, J. Mark MacNeil, JoAnn VanKessel, Bruce J. Forrest, Ned A. Porter, Laura S. Lehman, Karl J. Smith, Joe C. Ellington , *J. Am. Chem. Soc.,* 106. (48), 6740 (1984).

[21] Y.Chen, P. *Deuster: Chemico-Biological Interactions.* 182 (1), 7 (2009).

In: Chemical and Biochemical Kinetics: New Perspectives
Editor: Gennady Zaikov
ISBN 978-1-61209-224-9

Chapter 17

PHYSICOCHEMICAL AND ANTIMICROBIAL PROPERTIES OF NANOCOMPOSITE FILMS BASED ON HYDROXYETHYLCELLULOSE

Olga V. Alekseeva,* Svetlana A. Chulovskaya, Nadezhda A. Bagrovskaya and Vladimir I. Parfenyuk
Institution of Russian Academy of Sciences Institute of Solution Chemistry of the RAS, Ivanovo, Russia

ABSTRACT

Polymer hydroxyethylcellulose matrix modification was made by mechanochemical method using copper containing nanoparticles, obtained by electrochemical reduction from copper sulphate solutions in water- ethanol medium. Interactions of copper containing nanoparticles with hydroxyethylcellulose were studies. Antimicrobial effect of the nanocomposites obtained was found.

Keywords: Cellulose, Metal nanoparticles, Electosynthesis, Nanocomposites,Antimicrobial Properties

INTRODUCTION

Synthesis of nanocomposite materials and studies of their properties are the most dynamically developing directions of modern technologies [1-3]. Materials containing metal nanoparticles are extensively studied due to possible applications of their physicochemical properties unobserved for the corresponding bulk solids [4, 5].

Ultradispersed metal and metal oxide powders are often used for preparation of new polymer composites. This is because small size and high surface energy of the nanoparticles results in unique properties of immobilized materials such as magnetic, nonlinear optical,

* E-mail: avn@isc-ras.ru.

catalytic, etc. Use of copper containing powders for preparation of medical material with biocidic properties with respect to many kinds of bacteria is very perspective [6, 7].

However, synthesis of metal nanoparticles is connected with certain difficulties due to thermodynamic instability of the particles. The copper contained powders were prepared by cathode reduction method from electrolyte solutions in water-organic solvents [8, 9]. It should be noted that a high degree of dispersivity of electrodeposited particles can be reached in the presence of surfactants at limiting current densities [10]. The limiting current density provides the uniform delivery of metal cations to all points on the cathodic surface to compensate their loss at their discharge. Organic solvents exclude the interaction between the particles and to keeping of their optimum size characteristics and thermodynamic stability.

Methods of immobilization of nanoparticles in polymer matrix are classified as physical, physicochemical and chemical [11]. The physical method consists in mixing of dispersed particles and dispersing polymer medium. The advantage of this method in comparison with others is possibility of synthesis of the nanoparticles of given size which are stable during long time. As stabilizers, water-soluble cellulose derivatives can be used. These polymers immobilized by metal nanoparticles are antimicrobial and can be applied for development of various medical materials [12].

The aim of this work is the synthesis of the ultrafine copper containing powders by cathode deposition, their immobilization into cellulose matrix and the study of physicochemical and antimicrobial properties of the obtained nanocomposites.

EXPERIMENTAL

Synthesis of the ultrafine copper containing powders was carried out in glass electrochemical cell equipped with a steel cathode and two ruthenium-titanium oxide anodes (Figure 1). Aqueous solution of copper sulphate with additions of ethyl alcohol was used as electrolyte.

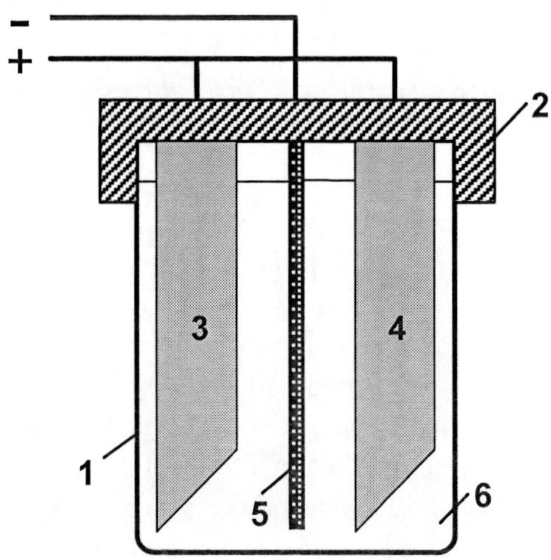

Figure 1. Schematic presentation of the experimental setup: electrochemical cell (1); vinyl lid(2); anodes (3,4); cathode (5); electrolyte solution (6).

The metal spongy deposit is formed on the cathode. After finishing of the process the spongy deposit was isolated from the cathode and was washed with distilled water up to constant electroconductivity of the mother solution. Then the deposit was dried at room temperature up to constant mass. The obtained powder was brown.

To determine the working electrolyte concentration and the current density corresponding to the limiting mass transfer mode, the first step was the polarization measurements carried(Figure 1). For a linear potential scanning (scanning rate 5 mV/s, $E = E_0 + a\tau$) due to the slowly changing conditions, a mode close to the steady state measurements was maintained. As the working electrode, a copper disk was used. A platinum electrode served as the counter electrode.

Hydroxyethylcellulose ($\{C_6H_7O_2(OH)_{3-x}[(OCH_2CH_2)_yOH]_x\}_n$) (HEC, brand «Hercules» U.S. with a molecular mass of ~ 300000) was chosen as polymer matrix for immobilization of the copper nanoparticles. The nanocomposite films were obtained by mechanochemical dispersion of the copper containing particles in HEC matrix. The aqueous solution of HEC (3 wt. %) was prepared and remained for swelling. Then, copper powder (0.035 wt. % of the polymer mass) was added. The film was prepared on glass support and was dried up to complete solvent removal. The obtained film (thickness of about 70 micrometers) was transparent and had reddish tint.

The morphology of powder and polymer compositewas determined by an EMV-100 transmission electron microscopy (TEM) with an accelerating voltage of 50-100 kV, resolution 3 Å and magnification of from 2000 to 50000. The prepared powder suspension in ethanol was sprayed on copper grids covered with carbon films using a supersonic dispenser nozzle.

IR spectra of the modified and unmodified polymer films were recorded by Avatar ESP-360 spectrometer in the range from 4000 to 600 cm^{-1}.

X-ray investigation was performed using a DRON-3M diffractometer using CuK_αradiation. The determination of the chemical composition was carried out by correlationof the parameters of the studied object's diffraction lines with reference data [13].

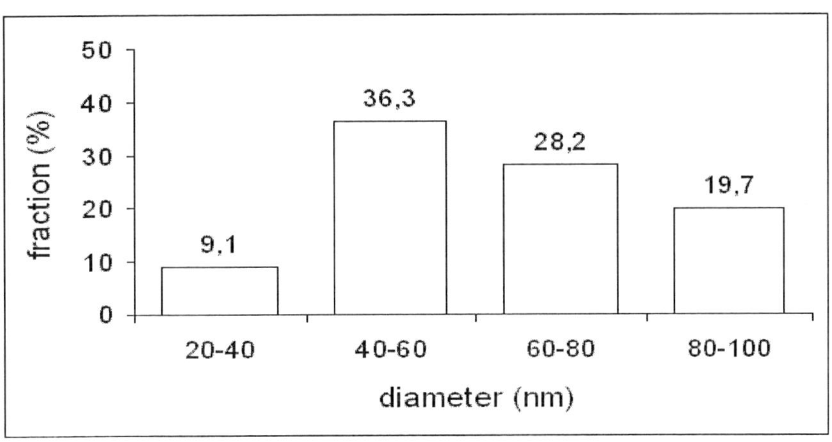

Figure 2. Size distribution of copper particles in the powder.

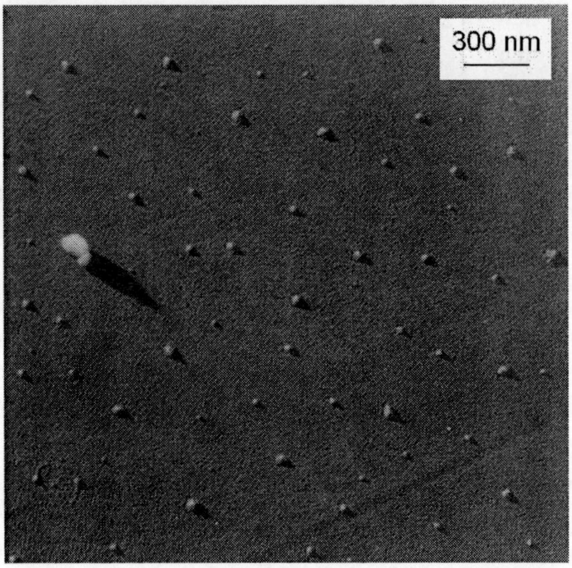

Figure 3. The TEM micrographs of the nanoparticles of the copper containing powder.

Antibacterial efficiency of the nanocomposites against colon bacillus and staphylococcus was checked by bacterial inoculation of "lawn" type in Petri dishes. 1, 5-2 ml of the bacterial dredge containing 500 millions microbial cells per ml was placed on the surface of beef-extract agar. Then above it the samples of the films under study were placed. The dredge was distributed evenly. Its excess was removed by pipette. The dishes were kept in thermostat for 24 h. Inhibition of growth of bacteria is criterion of the antibacterial efficacy.

DISCUSSION

Histogram showing the size distribution of the copper nanoparticles prepared from aqueous ethanol alcohol (0.1 mol·kg^{-1} $CuSO_4$, 0.04 mole fractions of the alcohol) is presented in Figure 2. The histograms were obtained directly from the TEM images.

As can be seen from Figure 2, ~93 % of the particles have the diameter from 20 to 100 nm and~7 % are larger aggregates (100-500 nm). The obtained powder contains also unoxidized copper, Cu (I) and Cu (II) oxides.

Figure 3 shows the TEM micrographs of the nanoparticles of the copper containing powder.

Histogram of size distribution of the copper particles in HEC matrix are presented in Figure 4. As can be seen, the particles of the nanocomposite have sizes in the range of 20-100 nm. Amount of the larger aggregates (100-500 nm) is not greater than 10 %.

Thus, the size distribution of the copper particles in the copper containig powder and in the HEC nanocomposite is slightly different. It is likely that the observed increase of the particle fraction having a size of 40-60 nm and decrease of the particle fraction having a size of 80-100 nm in HEC matrix as compared with the copper powder is associated with intensive mixing at preparation of the nanocomposite.

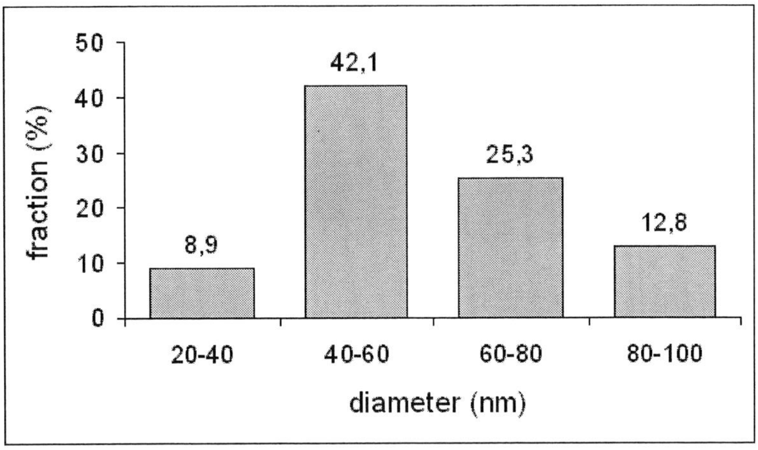

Figure 4. Histogram of size distribution of the copper particles in HEC matrix.

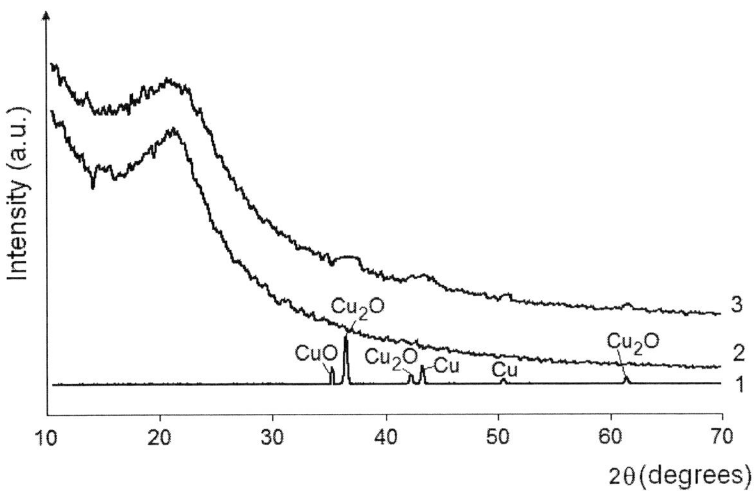

Figure 5. The X-ray diffraction patterns of nano-sized copper powder (1), the initial (2) and modified (3) HEC films.

X-ray diffraction can give us information about the structure of the studied objects. The diffraction patterns of nano-sized copper powder (1), the original (2) and modified (3) HEC filmsare shown Figure 5. The reflections of nano-sized powder on the diffraction pattern are observed. The most intense of them are ($2\theta^o$ = 35,73 (CuO); 36,63 (Cu_2O); 42,59 (Cu_2O); 43,90 (Cu); 50,35 (Cu); 61 , 32 (Cu_2O). Thus, the nano-sized copper powder is multidimensional and includes the unoxidized copper and copper oxides (I, II).

The diffraction pattern of the initial HEC film (2) contains only the amorphous halo. The center of its broad maximum is located at $2\theta \sim 22^o$. On the diffractogram of the modified HEC film (3) the narrowing main halo and new reflexes are observed. Their locations coincide with the reflections of nano-sized copper-containing powder. So it is likely thatthe introduction of nano-sized copper particles into the amorphous HEC film leads to changes in its structure and formation of crystalline areas.

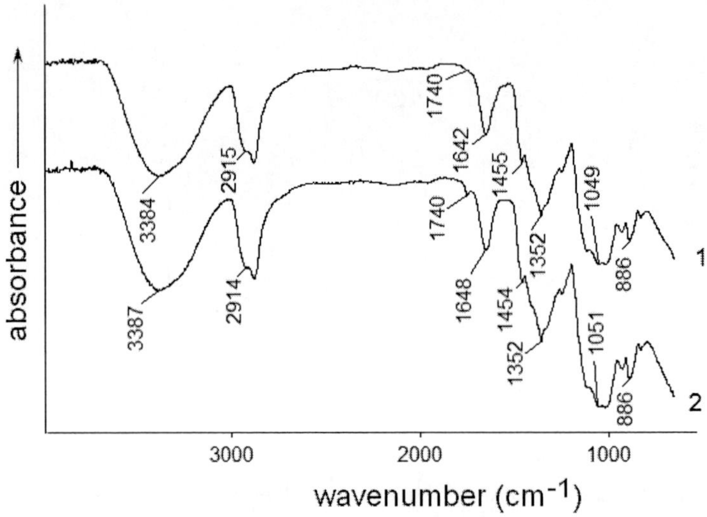

Figure 6. IR spectra of the unmodified (1) and modified (2) films of HEC.

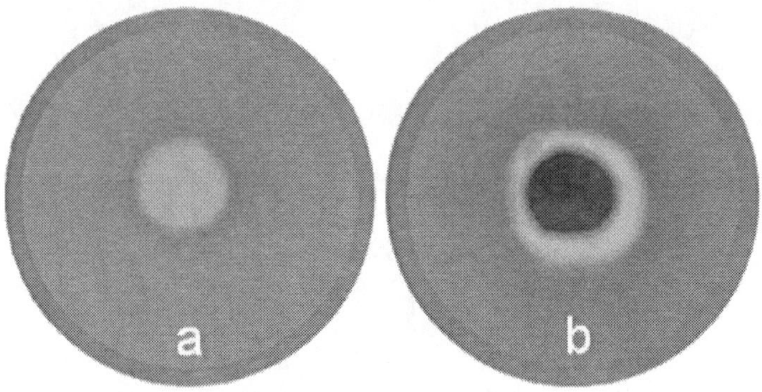

Figure 7. The effect of the unmodified (a) and the modified (b) films on staphylococcus.

Table. Infrared data of the unmodified and modified films of HEC

Vibration bands	Unmodified HEC			Modified HEC		
	ν, cm^{-1}	D	D/D_{827}	ν, cm^{-1}	D	D/D_{827}
$\nu_{O\text{-}H}$	3384	0.24	4.00	3387	0.21	3.42
$\nu_{C=O}$	1745	shoulder	-	1747	0.015	0.25
$\delta_{O\text{-}H}$	1352	0.32	5.33	1351	0.28	4.67
$\nu_{C\text{-}OH}$	1046	0.52	8.67	1053	0.44	7.33
$\delta_{C\text{-}H}$	827	0.06		827	0.06	

To obtain information about changes upon modification of HEC by the copper nanoparticles, IR spectra of the films obtained were recorded (Figure 6).

In the IR spectrum of the unmodified HEC a broad band at 3600-3000 cm^{-1} is assigned to stretching vibrations of H-bonded OH groups. A band at 2900 cm^{-1} is assigned to stretching vibrations of CH_2 and CH groups of the polymer. O-H deformation vibrations at 1352 cm^{-1} are a part of a broad band at 1450-1200 cm^{-1}. An intensive maximum at 1046 cm^{-1} is assigned to stretching C-OH vibrations. A band at 1642 cm^{-1} is associated with crystallization water. A shoulder at 1745 cm^{-1} indicates on a small amount of C=O groups in the polymer [14].

The IR spectra of the unmodified and modified films of HEC differ from each other. In the spectrum of the modified film the pronounced increase of intensity of the band at 1747 cm^{-1} is observed. The band at 1046 cm^{-1} exhibits a shift of 7 cm^{-1} to high-frequency region. It should be noted that the interpretation of these spectra is not an easy task because of superposition of some bands in different regions of the spectrum. In this work an attempt of semi-quantitative analysis of the IR spectra using method of base line and internal standard [15] was undertaken. A band at 827 cm^{-1} assigned to C-H vibrations was chosen as the internal standard. Ratios of transmission density of a series of the characteristic bands in the IR spectra of the films to the transmission density of the band at 827 cm^{-1} are listed in Table. In the spectrum of the modified film the relative intensities of the bands at 3384, 1352 and 1046 cm^{-1} decrease significantly. It is well known that copper oxides are efficient catalysts of oxidation of organic compounds. It is likely that the decrease of the relative intensities characterizing different modes of vibrations of O-H bond indicates on the oxidation of hydroxyl groups of the polymer and formation of C=O groups. This is confirmed by appearance of the weak $\nu_{C=O}$ band at 1745 cm^{-1}.

The results of the tests of antimicrobial efficiency of the obtained filmsagainst colon bacillus and staphylococcus showed absolute death of the bacteria under the modified film. It should be noted that dynamics of inactivation of the bacteria keeps during a month.

The effect of the unmodified (a) and the modified (b) films on staphylococcus is presented in Figure 7.

It is likely that one of the reasons of bacterium death is interaction of the metal nanoparticles with functional groups of the amino acids composing bacterial proteins. This results in the cell membrane damage and destruction of the cell wall leading their death.

CONCLUSIONS

Nano-sized copper containing powders were prepared by electrochemical cathode precipitation from solution of copper (II) sulphate in water- ethanol alcohol mixtures. The immobilization of the copper containing nanoparticles in hydroxyethyl cellulose matrix was realized. Size distribution of the copper containing nanoparticles in hydroxyethyl cellulose matrix as well as their interaction with the polymer was studied. It was found that the obtained nanocomposites exhibit antimicrobial activity against colon bacillus and staphylococcus.

REFERENCES

[1] Lysenko A.A., Astashkina O.V., Sverdlova N.I. et. al. Metall-containing chemical fibers and their utilizationin biotechnology. 2007. *Chemical Fibers.* 2: 44-50.

[2] Ławecka M., Kopcewicz M., Slawska-Waniewska A., Leonowiczl M., Kozubowski J., Dzhardimalieva G.I., Rozenberg A.S. and Pomogailo A.D.2003. Formation, structure and magnetic properties of polymer matrix nanocomposites processed by thermal decomposition of the Fe(III) Co(II) acrylate complex. *J. Nanoparticle Research.* 5:373-381.

[3] Ozerin A.N., Zelenetskii A.N., Akopova T.A. et. al. Nanocomposites based on modified chitosan and titanium oxide. 2006. *Polymer Science Series A.* 48: 638-643.

[4] Ramos J., Millan A. and Palacio F. 2000. Production of magnetic nanoparticles in a polyvinylpyridine matrix. *Polymer.* 41: 8461-8464

[5] You T.Y., Niwa O., Tomita M. et. al. 2002. Characterization and electrochemical properties of highly dispersed copper oxide/hydroxide nanoparticles in graphitelike carbon films prepared by RF sputtering method. *Electrochem. Commun.* 4: 468-471.

[6] Mahapatra O., Bhagat M., Gopalakrishnan and Arunachalam Kantha B. 2008. Ultrafine dispersed CuO nanoparticles and their antibacterial activity. *J. of Experimental Nanoscience.* 3: 185-193.

[7] Elinson V.M., Sleptsov V.V., Potraysay V.V., Laymin A.N. 2004. *Biomedical properties of nanostructured surfaces formed by ion-plasma methods* // NATO, Science Series, Cluver Publisher, USA, P. 307-319.

[8] Chulovskaya S.A., Balmasov A.V., Lilin S.A. and Parfenyuk V.I. 2006. Electrochemical Production of Ultradisperse Copper-Containing Particles from Organo-Aqueous Electrolyte Solutions. *Protection of Metals.* 42: 394-397

[9] Sinha A., Das S.K., Kumar T.V.V., Rao V. and Ramachandrarao P.J. 1999. Syntesis of Nanosized Copper Powder by an Aqueous Route. *Mat. Synth. Proc.* 7: 373-377.

[10] Chulovskaya S.A. and Parfenyuk V.I. 2007. Influence of the isopropyl alcohol on the dimensional and qualitative composition of ultradispersed copper-contained powders *Russian J. Appl. Chem.,* 80: 952-955.

[11] Pomogailo A.D., Rozenberg A.C., Uflyand I.E., in Nanochastitsy metallov v polymerakh, Moskva, Khimiya, 2000,s. 325–331. [Pomogailo A.D., Rozenberg A.C. and Uflyand I.E., in Nanoparticles of metals in polymers, Moscow, *Chemistry,* 2000, ch. 4, pp. 325–331.] [In Russian].

[12] Elinson V.M., Sleptsov V.V., Potraysay V.V., Laymin A.N. 2004. *Biomedical properties of nanostructured surfaces formed by ion-plasma methods* // NATO, Science Series, Cluver Publisher, USA, P. 307-319. // Tomsk

[13] Mirkin L.I. Spravochnik po rentgenostrukturnomu analizu polikristallov(*Handbook on X-ray Structural Analysis of Polycrystals*), Moscow: Fizmatizdat, 1961.

[14] Blackwell J., Marchessault R.H., *in Cellulose and cellulose derivatives,* ed. N.M. Bikales and L. Segal. 1971. Wiley-Interscience, New York, part IV, ch.1, pp.3-18.

[15] Smith A.L., in Applied infrared spectroscopy. *Fundamentals, techniques, and analytical problem-solving. New* York: Wiley. 1979.

In: Chemical and Biochemical Kinetics: New Perspectives ISBN 978-1-61209-224-9
Editor: Gennady Zaikov

Chapter 18

SEVILENE-BASED HYBRID COMPOSITES

O. A. Legonkova,[1,*] A. A. Popov,[2] A. A. Bokarev,[1] and S. G. Karpova[2]

[1]Moscow State University of Applied Biotechnology, Russia
[2]Emanuel Institute of Biochemical Physics, Russian Academy of Sciences, Russia

ABSTRACT

The work presents the results of developing hybrid high-filled composites with regulated physico-mechanical properties based on sevilene (SEVA) of grades 11306-070, 11506-070 and on organic and inorganic fillers. The process properties of hybrid composites are related to not only the decrease of the concentration of the matrix in the bulk of the composite, but also to the plasticizing action of inorganic filler.

Keywords: hybrid composites, utilization,industrial wastes, polymers, physical and chemical properties, structure, technological properties, organic fillers, inorganic fillers, correlation time, melt flow index

AIMS AND BACKGROUNDS

The aim of the carried out work was to elaborate the highly filled composite materials with regulated physical and mechanical properties based on synthetic polymers and organic and inorganic fillers that can be used in creation of biodegradable in the environment goods of different purposes. Aspects, essential for technology, such as physical and chemical, structural, rheological properties: were investigated. Methods of electron paramagnetic probe, IR-spectrometry were used to prove that inorganic filler plays the role of plasticizer in creation of hybrid composites. Microbiological aspects will be the subject of another article.

* E-mail: OALegonkovaPB@mail.ru.

INTRODUCTION

One of the promising trends from the viewpoint of ecology is the development of biodegradable polymer composites. These materials, along with the polymer base more resistant to biodegradation, comprise fillers, which are not only accessible for microbial degraders but are also agro-industrial wastes to be utilized [1]. Search for cheap fillers and development of polymer composites makes it possible not only to reduce the cost of product, but also contributes to the solution of ecological problems.

EXPERIMENTAL PART

Third-grade threshed grain wastes (size of particles, 63–240 μm; bulk density, 350 kg/m^3; humidity, 4%) were used as organic filler. As inorganic filler, we took a Rastvorin-A water-soluble mineral fertilizer (OST 10-193-96, produced by the Buysk Mineral Fertilizer Plant) of the following composition (in %): $(NH_4)_2SO_4$, 35; $NH_4H_2PO_4$, 6; KNO_3, 32; $MgSO_4 \cdot 7H_2O$, 27.

The physico-mechanical properties of specimens within a broad range of ingredient ratios were determined according to GOST 14236-82; the rheological characteristics of filled compositions, by the method of capillary viscosimetry.

In the work, use was made of the method of electron paramagnetic probe, which was stable nitroxyl radical 2,2,6,6-tetramethylpiperidine-1-oxyl. The radical was introduced into films from vapours at $T = 25°C$ up to a concentration of 10^{-3} mol/l. A reference solution of the radical in CCl_4 with a known number of spins was used in the determination of the concentration of radicals in films. The number of spins in a specimen was determined by comparing the areas under the absorption curves of the specimen studied and the reference. The rotational mobility of the probe was characterized by the correlation time τ. The values of τ were assessed from the EPR spectra by the method outlined in [2].

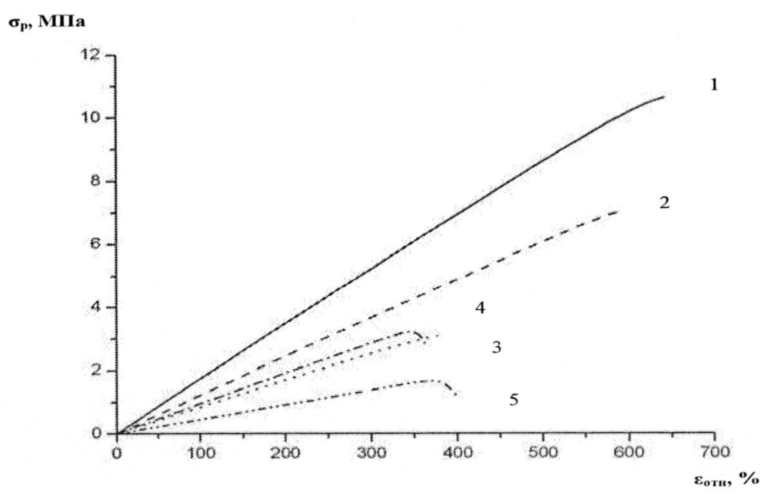

Figure 1. Change of the physico-mechanical properties of a two-component system (SEVA / inorganic filler, wt. %): 1, 100/0; 2, 80/20; 3, 60/40; 4, 50/50; 5, 40/60.

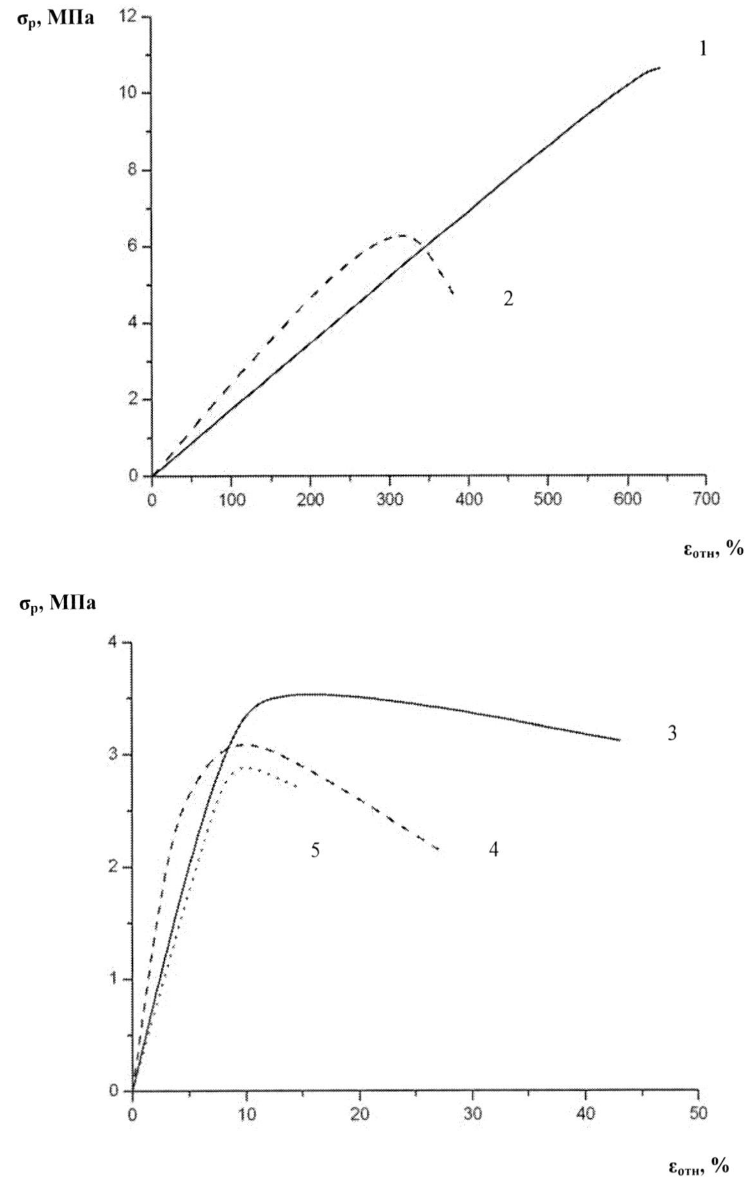

Figure 2. Change of the physico-mechanical properties of a two-component system (SEVA / organic filler, wt. %): 1, 100/0; 2, 80/20; 3, 60/40; 4, 50/50; 5, 40/60.

Results and Discussions

At the introduction of fillers into SEVA irrespective of its grade, we found that the strength and relative elongation decrease with an increase of the content both of organic and inorganic filler in two-component polymer–filler systems, see Figures 1 and 2. In the case of filling with organic filler specimens become more rigid; at the introduction of inorganic filler even at a high concentration (60 wt. %) specimens preserve a high plasticity (the breaking strain is 400%).

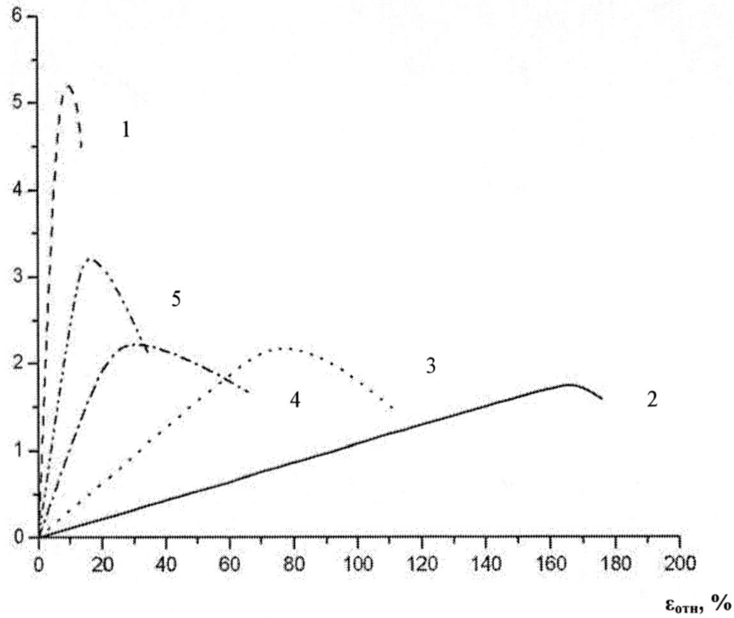

Figure 3. The physico-mechanical properties of a three-component system based on SEVA-113, wt. %: 1, SEVA/organic filler, 50/50; 2, SEVA/inorganic filler, 50/50; 3, SEVA/inorganic filler/organic filler, 50/37/13; 4, SEVA/inorganic filler/organic filler, 50/25/25; 5, SEVA/inorganic filler/organic filler, 50/13/37.

Considering three-component systems (SEVA/inorganic filler/organic filler, Figure 3), it should be noted that in this case too an increase in the content of inorganic filler leads to more plastic specimens.

Figure 4 presents electron micrographs of the fracture faces of composites. The systems are seen to be heterogeneous. The systems with inorganic filler include both crystallites of filler salts and their more complex formations. Defects in the system are “pressed into” the homogeneous structure of the composite. For this reason, a decrease of strength of polymer–composite material occurs owing to a decrease of the content of polymer in the composition.

The EPR method supports the data obtained (Table 1): the higher the concentration of organic wastes in a hybrid composite, the greater the correlation time, i.e., the more rigid the system is and the smaller the deformability of the system is. At the same time, the higher the concentration of inorganic filler, the more plastic the systems are.

Noteworthy is the behaviour of two ternary systems: 50% sevilene at 25/25 of organic and inorganic filler; and 15% sevilene at 50/35 of organic and inorganic filler, for which the correlation time of all ternary systems studied is minimal. The systems are similarly plastic and similarly capable of being processed by commercial equipment into particular articles.

Given that inorganic filler is a mixture of salts (potassium, magnesium chlorides; magnesium sulfates; potassium and ammonium nitrates), it was shown (Table 2) that exactly nitrogen-containing salts led to a significant increase of the melt flow index in sevilene-based PCMs.

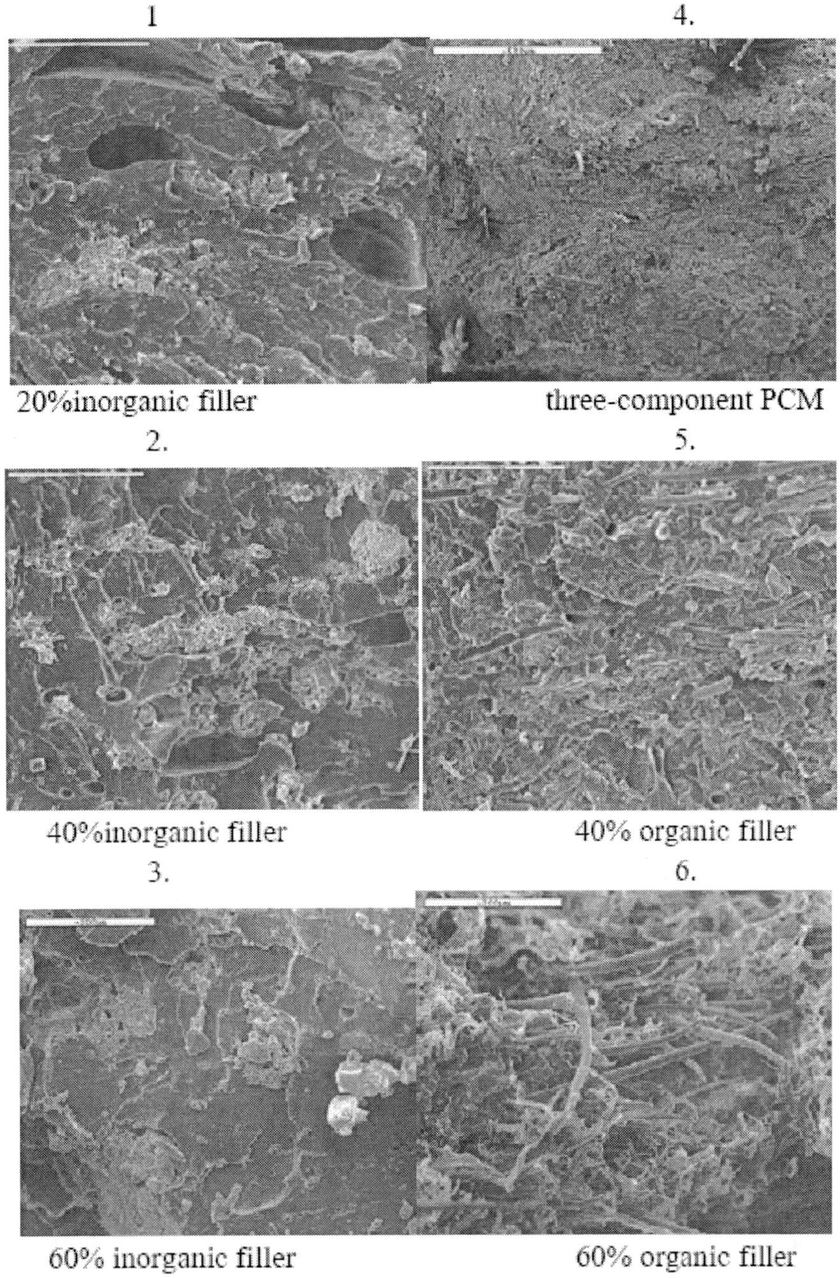

Figure 4. Electron micrographs of the fracture faces of two-component systems of polymer–composite material based on sevilene and inorganic filler (1, 2, 3); sevilene and organic filler (5, 6); three-component PCM (4): 15% sevilene, 50% organic filler, 35% inorganic filler (magnification ×200 μm).

To assess the possibility of chemical interaction of fillers with polymers, we used the method of IR spectroscopy. The following IR bands were found to be characteristic of SEVA: 1740 cm^{-1} (vibrations of the aldehyde group), 1380 cm^{-1} (vibrations of the vinyl acetate group), 1240 cm^{-1}, 1050 cm^{-1}, 797 cm^{-1}, 607 cm^{-1} [3].

Table 1. Correlation time of the paramagnetic probe in PCM specimens depending on the extent of filling with various fillers

Extent of filling with organic filler, %	τ	Extent of filling with inorganic filler,%	τ	Extent of filling with organic/inorganic filler, %	τ
10	$6.1\cdot10^{-10}$	20	$7.7\cdot10^{-10}$	18.75/18.75	$6.1\cdot10^{-10}$
15	$6.8\cdot10^{-10}$	40	$7.4\cdot10^{-10}$	25/25	$5.1\cdot10^{-10}$
20	$7.1\cdot10^{-10}$	50	$7.1\cdot10^{-10}$	50/25	$6.2\cdot10^{-10}$
35	$6.3\cdot10^{-10}$	60	$6.8\cdot10^{-10}$	60/25	$6.5\cdot10^{-10}$
40	$7.0\cdot10^{-10}$	80	$6.4\cdot10^{-10}$	30/35	$6.0\cdot10^{-10}$
50	$7.1\cdot10^{-10}$			50/35	$6.2\cdot10^{-10}$
60	$8.0\cdot10^{-10}$			50/35	$5.7\cdot10^{-10}$

Table 2. Effect of the composition and concentration of inorganic salts on the melt flow index (g/10 min) of sevilene-based composites

Content, wt. %	Unfilled specimen	$MgCl_2$	KCl	$MgSO_4$	KNO_3	NH_4NO_3
1	7.8	7.5	6.7	8.1	2.9	4.3
2	7.5	7.1	6.3	7.1	Too high flow rate made its measurement impossible	

Table 3. Ratios of the optical densities of the bands 1440 cm^{-1} and 1380 cm^{-1} to the optical density of the band 2850 cm^{-1} taken as the internal standard of the specimens of mixtures of SEVA with the salts KNO_3, NH_4NO_3 and K_2HPO_4

	Sevilene	Sevilene $+KNO_3$	Sevilene $+NH_4NO_3$	Sevilene + K_2HPO_4	Sevilene + inorganic filler
D_{1440}/D_{2850}	0.66	0.47	0.46	0.42	0.41
D_{1380}/D_{2850}	0.59	0.42	0.43	0.31	0.36

Table 4.Melting heats of polymers and their composites at temperatures of solid-phase transitions of nitric salts (J/g)

Material	80°C	130°C
Sevilene unfilled (T_t = 80°C)	49.9	–
KNO_3	–	54.2
NH_4NO_3	10.1	53.9
Sevilene filled with 20% NH_4NO_3	49.9	9.8
Sevilene filled with 20% KNO_3	44.9	7.6
Sevilene filled with 2% KNO_3	65.1	1.1
Sevilene filled with 40% inorganic filler	50.4	
Sevilene/organic filler/inorganic filler (15:50:35)		94.9

According to [3], vibrations in the region of 1440 cm^{-1} can be assigned to –C—
|
$OCOCH_3$,

the methylene groups $-CH_2-$ in SEVA have characteristic valent vibrations at 2850 cm^{-1}. The group NH_4^+ has the absorption band in the region of 3300–3030 cm^{-1}; the group P-O-C_{alkyl}, 1180-1150 cm^{-1}; the groups SO_4^{2-}, 1130–1080 cm^{-1}; S-CH_2, 2700–2630 cm^{-1}; the groups PO_4^{3}, HPO_4^{2-}, $H_2PO_4^{-}$, in the region of 1100-1000 cm^{-1}; NO_3^{-}, 1380 cm^{-1}.

Changes in the ratio of the optical density of the –C($OCOCH_3$)– (D_{1440}) band, of the optical density of the vibration bands of the vinyl acetate group (D_{1380}) to the optical density of the bands, assigned to vibrations of the methylene groups (D_{2850}), were used to assess the changes in the composition of polymer at temperatures of producing PCMs together with salts (Table 3). It follows from the table that at a heating with the salts KNO_3, NH_4NO_3, K_2HPO_4 there is an insignificant splitting-off of the vinyl acetate group from the main chain. Heating of polymer with K_2HPO_4 produces a band responsible for the interaction of P-O-C_{alkyl} (D_{1180} = 0.62).

In the case of the other salts (KH_2PO_4, $(NH_4)_2HPO_4$,$(NH_4)_2SO_4$,$MgSO_4$), such an effect could not be identified due to the coincidence of the spectra of the characteristic absorption bands.

Table 4 presents the melting heats of polymers and polymer–composite materials at temperatures of solid-phase transitions of nitric salts, coinciding with sevilene fluidity temperature of 80°C. According to [4], ammonium nitrates have solid-phase transitions at temperatures of 80°C and 130°C with ΔH_m equal to 0.32 ccal/mol and 1.01 ccal/mol, respectively; the salt KNO_3 has the solid-phase transition at 130°C with ΔH_m equal to 1.3 ccal/mol.

At 80°C, for polymer–composite material based on sevilene an increase of melting heat at the introduction of 2% KNO_3 is due to the fact that densely packed structures are formed in polymer's liquid phase at a small content of solid filler at the interface with solidparticles of filler KNO_3 [5]. At an increase of the content of salt up to 20% the decrease of the melting heat is due to a decrease of the content of polymer per 1 g of specimen studied.

At the introduction of NH_4NO_3 into this system, there is no change of melting enthalpy as compared with pure polymer. At 130°C, when the temperature of the solid-phase transition of the salts is higher than the melting temperature of sevilene, the enthalpy of the composite decreases, which is due to a decrease of the content of salt per 1 g of specimen studied.

CONCLUSIONS

Thus, the thermal effect of two-component hybrid systems formed from the melt is minimal when T_m (T_t) of the polymer is equal to or lower than the solid-phase transition temperature of the salts. By the extent of a decrease of enthalpy of the measured component, we can make an indirect judgement on the looseness of the filled system as compared with the individual components of the composite.

The process properties of hybrid composites are related to not only the decrease of the concentration of the matrix in the bulk of the composite, but also to the plasticizing action of inorganic filler.

REFERENCES

[1] Legonkova O.A. "Biodeterioration of Polymer-based Composite Materials in the Environment." *Materials Science*, 2008, 2, 50–55.

[2] Antsyferova L.I., Wasserman A.M., Ivanova A.I., Lifshits V.A., Nazemets N.S. "*Atlas of the Spectra*" , Moscow: Nauka Publishers, 1977, 160 pp. (in Russian).

[3] Nakanishi K. *Infrared Spectra and Structure of Organic Compounds,* Moscow: Mir Publishers, 1965, 200 pp.

[4] *Chemist's Reference Book Moscow–Leningrad,* Khimiya Publishers, 1966, vol. 1, 1070 pp. (in Russian).

[5] Bryk M.T. "*Utilization of Filled Polymers",* Moscow: Khimiya Publishers, 1989, 190 pp. (in Russian).

In: Chemical and Biochemical Kinetics: New Perspectives
Editor: Gennady Zaikov
ISBN 978-1-61209-224-9

Chapter 19

EFFECT OF CHITOSAN MOLECULAR WEIGHT ON RHEOLOGICAL BEHAVIOR OF CHITOSAN MODIFIED NANOCLAY IN HIGHLY HYDRATED STATES

Linshu Liu,[1,*] Songmiao Liang,[2] Liwen Chen,[2] Peng Yuan,[2] Qingrong Huang,[2] and Kit L. Yam[2]

[1]Eastern Regional Research Center,
United States Department of Agriculture, Wyndmoor, PA, U. S. A.
[2]Department of Food Science, Rutgers University,
New Brunswick, NJ, U. S. A.

ABSTRACT

Effect of chitosan molecular weight (M_{cs}) on the rheological properties of chitosan modified clay (CMCs) at highly hydrated state was investigated. With special emphasis on its effect on thethixotropy of CMCs, the structure recovery at rest after underwent a pre-shearing process was further performed by the oscillatory shear measurements in the linear viscoelastic region. It wasobserved that both stress-strain behavior and moduli of CMCs are closely correlated to M_{cs}. An interesting slope transition from negative to positive in the tan δ vs log ω plot is observed at $\omega = 2.5$ rad/s. The value of $\tan \delta < 0.5$ over the entire range of frequency suggests a strong association in CMCs and a dominant elastic response of CMCs. Small peak-like transition of the shear viscosity, which denotes the yield region of the CMCs network, shifts to high shear rate with the increase of M_{cs}. Remarkably, different thixotropic behavior of these considered CMCsare observed relating to M_{cs}. G'_{After} of CMCs with high M_{cs} shows a linearly monotonic increase with the increase of the testing time, which is well match the power law of $G'_{After} \sim t^n$. G'_{After} of those CMCs with low M_{cs}, however, is distinctly deviated from the power law and possesses an exponential increase. The deviation of G'_{After} becomes more marked with the decrease of M_{cs}. Furthermore, the effect of the pre-shearing history on the thixotropic properties of CMCs is performed by employing a large range of the pre-shear rate. Analysis of

* Correspondening author. Tel. +1(215)518-1623; Email Address: Linshu.liu@ars.usda.govMention of brand or firm names does not constitute an endorsement by the U.S. Department of Agriculture above others of a similar nature not mentioned.

empirical Cox-Merz rule indicates a failure of the rule in CMCs system. CMCs prepared from low M_{cs} chitosan are easier to be orientated at low shear rate.

Keywords: Chitosan, Molecular weight, Clay, Hydrated state, Rheological behavior

INTRODUCTION

As one of the most commonly used inorganic fillers, silicate clay has been widely applied todesign hybrid materials with promising mechanical and thermal performance.[1-7] Silicate claypossesses a highly layered and anisotropic structure with the layer thickness of about 1 nm and the lateral dimension between 100 to 1000 nm.[8, 9] Negatively charged sites usually exist on thesurface of each layer as a result of the replacement of $Mg^{2+}$to a part of Al^{3+} and Si^{4+} inaluminum-oxygen octahedron and silicon-oxygen tetrahedron of clay crystal cell. Consequently, hydrated sodium ions and other ions were absorbed into the interlayer space through interlayeradsorption to keep an electrostatic balance on its surface. However, these hydrated ions are notpermanently fixed and have the ability easily to exchange with other organic and inorganic ionsthrough cationic exchange process. [10-12] The ion exchange capability of silicate clay has beendetermined around at 0.9 ~ 1.2 mequiv./g.[10, 11] It has been reported that modification of clay with organic cations can not only effectively decrease its surface tension but promote its compatibilitywith polymer substrate and monomers.[13-18]More important, this kind of modification on clayprovides an essential way to introduce different functional groups so that design materials with required structure and performance. [19-22]

Developing special nanocomposites based on the interaction between clay and biopolymerssuch as proteins [23] and polysaccharides [24-27] has emerged as a hot topic in recent. For example, because of the polycationic nature of chitosan in acidic media, it has been reported that chitosancould intercalate into pristine clay to form chitosan/clay nanocomposites (CCNs) via cationicexchange process. [28-31] Bi-layered structure of chitosan has been built in the gallery space due tothe electrostatic interaction between clay and chitosan as well as the hydrogen bonding interaction between chitosan chains. CCNs with this special architecture has been explored to potentiometrically determine some anions, especially monovalent anions.[31] Moreover, enhancedmechanical properties and thermal stability of CCNs have been reported as a result of the strongclay-chitosan interaction [30] and the synergistic effect between 2D clay platelets and 1D carbonnanotubes (CNTs) in CCNs.[32] To our knowledge, however; there have been very few experimental exploration focusing on the rheological behavior of CCNs or CMCs at molten and highly hydrated state.

Very recently, we have successfully prepared CMCs with pre-exfoliated clay by means ofelectrostatic adsorption process. Microstructure and unique rheological behavior of CMCs at highly hydrated state have been studied as a function of the adsorbed chitosancontent. [33] As a continuation to our previous work, the present work would explore in detail the effects of M_{cs}on the rheological behavior of the highly hydrated CMCs by using oscillatory shear and steady-stateshear measurements. In general, the structure recovery at rest after steady shear was considered a fundamental thixotropic process. [34, 35] The effect of M_{cs}on their thixotropic properties is thereforeaddressed as a special emphasis of this work through the structure recovery of CMCs at rest afterunderwent a pre-shearing process. In addition, the adsorbed chitosan content and microstructureof CMCs at highly hydrated state are determined by thermogravimetric analysis (TG) and opticalmicroscopy, respectively. Our results clarified that the molecular weight of chitosan plays animportant role on both rheological properties and recovery behavior of the highly hydratedCMCs.

EXPERIMENTAL

Materials

Chitosan samples with deacetylation degree of98.0% and molecular weight (M_w) of44, 130 and330 KD were purchased from Kunpoong Bio. Co., LTD (South Korea). Silicate clay (Cloisite Na^+) was purchased from Southern Clay Products, Inc. All of thesereagents were used as received. Milli-Q (18.3 MΩ) water was used in all experiments.

Sample Preparation

Two wt% chitosan solutions with different molecular weight were obtained by dissolving 4g chitosan powder in 196g1 wt% acetic acid solution at room temperature. Fully exfoliated clay suspension was prepared according to our previous work. [33]5g Cloisite $Na^+$was dispersed in 495g distilled water by a magnetic stirrer(Thermix Stirring Hot Plate Model 210T, Fisher Scientific) at 1500 rpm for 72 h. The resultant solution was centrifugedat3000 rpm for 30 min to remove larger particles. The final concentration of fully exfoliated clay in the semitransparent suspension is 0.92 wt%.

CMCs were prepared from solution mixing method at room temperature. In this method, 3g o f2 wt% chitosan solutions with chitosan molecular weight of44,130,330 KD were separatelyinjected into20g 0.92 wt% clay suspensions by dropwise under a magnetic stirring of200 rpm.In all of the cases, the resultant mixtures were keeping stirring at200 rpm for1 h to establish adsorption equilibrium. After that, CMCs slurries prepared from the above molecular weight chitosan were obtained by a centrifugation of 3000 rpm for10 min. All of them were washed with pure water until free from acetate and were then sealed in centrifuge tubes for the next experiments.The water content of these CMCs slurries was measured as 97.1±1.2wt%. Concisely, CMCs prepared here were denoted as CMCs-x, where x=44,130 and330 arecorresponding to the chitosan molecular weight of 44, 130 and 330 KD, respectively.

Topography Analysis

The microstructure of CMCs at highly hydrated state was analyzed by Nikon TE-2000U inverted optical microscope with ×40 objective lens.

Thermogravimetric Analysis

TG analysis was employed to quantitatively determine thereal adsorbed content of chitosan in CMCs. Prior to the measurement CMCs slurries prepared from different molecular weight chitosan were dried in a vacuum oven at 50°C for 24 h and ground to powder. With sample weights of 10 mg, TG traces were then measured by a thermalanalysis system (TA, USA) in the temperature range of 300 ~ 1200 K under an air atmosphere.The adsorbed chitosan contents in the considered CMCs were approximately calculated from theweight loss of the samples at 500 ~ 800 K according to the previous work reported by M. Darderet al.[31] To remove the effect of water, the real contents of chitosan in the measured samples were calculated as:

$$M = \Phi_{cs} /(1- \Phi_{w}) \times 100\% \qquad (1)$$

Here, Φ_{cs} and Φ_{w} are the weight losses caused by chitosan and water, respectively.

Rheological Analysis

Oscillatory shear measurements of the highly hydrated CMCsslurries were performed on a strain-controlled rheometer (ARES, TA Instruments, New Castle, U.S.). Measurements were carried out using parallel plate geometry with 25mm diameter and 1mm gap at 25°C. Prior to measurement, the samples were loaded onto the plate and left for 20min to relax the stresses and obtain thermal equilibrium. Strain sweeps ranging from 0 to 100%were carried out to determine the proper conditions for dynamic measurements. Dynamic frequency sweep experiments measuring the complex viscosity (η*), storage modulus (G') andloss modulus (G'') were performed as functions of angular frequency (ω) ranging from 0.1 to100 rad/s. A fixed strain of 0.3% was used to ensure that measurements were taken well withinthe linear viscoelastic region. Steady-state shear measurements for CMCs slurries wereperformed over the shear rate range of 0.1~100 s^{-1}. The experimental protocols employed forthese measurements are virtually the same as those for the oscillatory shear measurements.Furthermore, to explore the effect of chitosan molecular weight on the thixotropic properties of CMCs, the samples were firstly loaded and pre-sheared under the shear rates of 0.1, 1, 10, 30, 50 and 100 s^{-1} for 300 s. Then, dynamic time sweep with an angular frequency of 1 rad/s and a strainof 0.3 % was applied immediately to record the recovery of G' and G'' within the testing time of1000 s. As a contrast, the same dynamic time sweep was also performed for the samples withoutpre-shearing treatment. However, rheological measurements for pure exfoliated clay were notsuccessfully performed here due to a failure to obtain clay slurry with the same water contentfrom its suspension. Combined with the literatures, it would not be the problem for us to make acomparison among the rheological properties of CMCs. For all of the experiments, a smallamount of low-viscosity silicon oil was coated around the edge of the sample to prevent waterfrom evaporation. The effect of the silicon oil on the overall rheological behavior is negligible.

RESULTS AND DISCUSSION

Effect of Chitosan Molecular Weight on the Microstructure of CMCs at Highly Hydrated State

To obtain high dispersion of clay in water and high electrostatic adsorption efficiency during the preparation process, the clay we used here was exfoliated in advance understrong mechanical stirring according to our previous work.[33] This treatment to pristine clay hasbeen successfully applied to prepare hybrid hydrogels, which have associative network andexcellent mechanical performance.[36-38] The adsorbed chitosan content were determined at 7.97,8.11 and 8.08 wt% for CMCs-44, CMCs-130 and CMCs-330, respectively. It shows no obvious dependence on chitosan molecular weight. This result implies that molecular weight of chitosan plays less effect on the finally adsorbed amount of chitosan on the clay surface. Since rheologicalproperties of materials are mainly determined by its time-dependence microstructure, the effectof M_{cs} on the microstructure of CMCs at highly hydrated state is further studied and shown inFigure 1. Network structure is observed for all of the samples. Moreover, with the increase ofchitosan molecular weight, this kind of network structure becomes more clear-cut. This changein the network structure is similar to that was observed in our previous work, which denoted astructure evolution from smooth to network structure of CMCs with the increase of the adsorbedchitosan content.[33] The structure evolution in the present case could arise due to the increase of chitosan chain length.

Effect of Chitosan Molecular Weight on the Dynamic Rheological Properties of CMCsat Highly Hydrated State

Figure 2 presents the applied strain (γ) dependence of G' and G'' atthe angular frequency of 1 rad/s. Both G' and G'' of CMCs increases with the increase of M_{cs} over the entire range of the applied strain. Payne effect, which deals with the strain-dependence of the dynamic viscoelastic properties of filled polymers in the amorphous state, is observed inthe considered CMCs. Moreover, in comparison with the strain-hardening behavior of the end-tethered polymer nanocomposites, [39] CMCs here exhibit a strain-softening response at most of theapplied strain range. A linear region of G' is observed at the strain region of $\gamma \leq 0.4\%$ while its nonlinear strain-dependence starts to appear at $\gamma > 0.4\%$. The range of the linear region of G' is independent of M_{cs}. However, the strain-dependence behavior of G'' looks more complicated ascompared to that of G'. It shows an interesting dependence on M_{cs}. There is a similarly quick decrease in G'' of CMCs-44, CMCs-130 and CMCs-330 at the applied strain of $\gamma \leq 0.25\%$,which possibly arises from the breakage of some interparticle structure formed by the weak interactions such as Van der Waal force and hydrogen bonding interaction. With a further increase of the applied strain, no strictly linear strain-dependence of G'' appears over the entire range of the applied strain for CMCs-130 and CMCs-330. G'' of CMCs-330 has the most obvious protuberance-like distribution at the applied strain range of $0.25\% \leq \gamma \leq 15\%$. It becomes a little slight for CMCs-130. G'' of CMCs-44 in this region is almost kept as a plateau. Regarding the difference in chitosan molecular weight, the protuberance-like distribution of G''at high M_{cs} possibly stems from the strong viscous response of chitosan with long chain length. Furthermore, crossovers between G' and G'' of CMCs are observed at the applied strain of γ >42 %, indicating a transition from solid-like behavior to liquid-like behavior of CMCs in thisregion. The critical strain (γ_c) for the appearance of the crossover shifts from about 42.0% to 54.0% with the increase of M_{cs} from 44 KD to 330KD. It implies that the network structure of CMCs formed by high M_{cs}chitosan is more stable against the external strain.

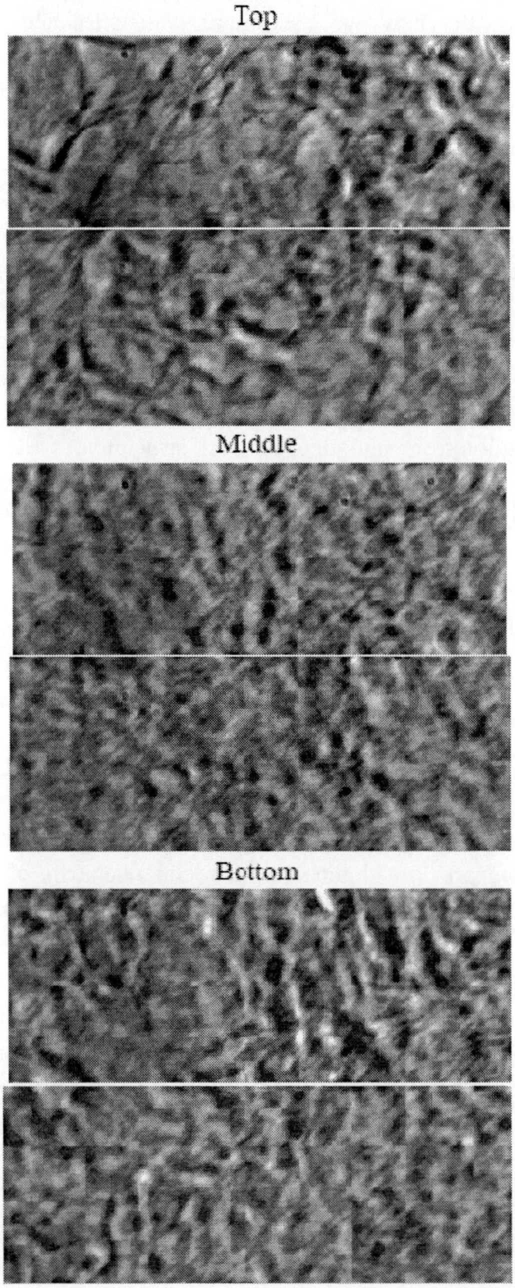

Figure 1. Optical microscopy images of CMCs-44 (Top), CMCs-130 (Middle) and CMCs-330 (bottom).

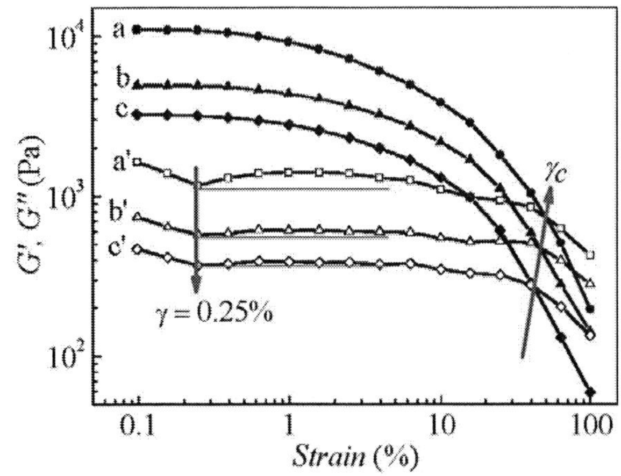

Figure 2. *G'* (a, b and c) and *G''* (a', b' and c') of CMCs-330 (a, a'), CMCs-130 (b, b') and CMCs-44 (c, c') as a function of the applied strain.

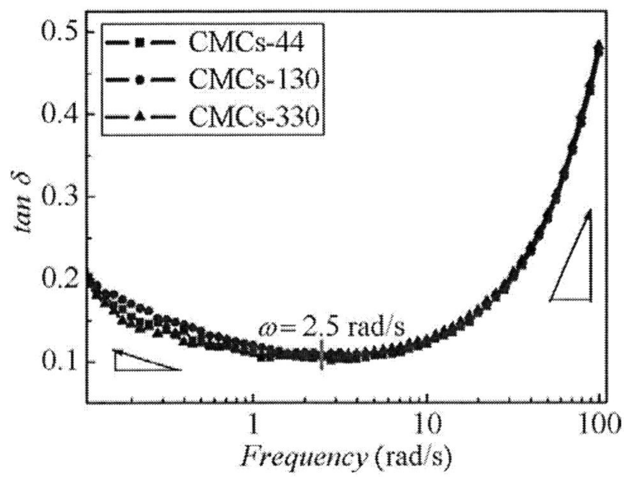

Figure 3. tan δ of CMCs as a function of the applied angular frequency at $\gamma = 0.3\%$.

Loss tangent (tan δ = G'' / G') depicted to the applied frequency is presented in Figure 3. Thevalues of tan δ are far smaller than unit over the entire range of the applied frequency, indicativeof an elastic-like behavior of CMCs. Interestingly, there is a transition in the slope of the tan δ vslog ω plot from negative to positive for all of the samples with the increase of the frequency. The critical frequency for the transition is approximately located at ω = 2.5 rad/s, which is independent of chitosan molecular weight. This special transition in the slope has been used todistinguish the different state of materials. Namely, for polymer melts or liquids, the slopes ofcurves in a tan δ vs log ω plot are negative while for the solids (gels) it is positive. [40, 41] However, this judgment is not applicable in the present case since no typical transition between elastic-liketo viscous-like of CMCs occurs over the entire range of the applied frequency. It likely arises from the breakage of some weak structure and/or the partial orientation of CMCs particles, which results in a quick increase in the viscous-like structure of CMCs. Furthermore, the amplitude of tan δ has found to be related to the intensity of the interaction between componentsin nanocomposites. Loss tangent δ > 3 indicates nonassociated; while, 1 < tan δ < 3 denotesweakly associated in the system. The strongly associated interaction is possessed when tan $\delta <1$. [42] tan δ for all of the considered CMCs is located at the range of 0.1 ~ 0.5, indicative of astrong association between chitosan and clay. Small deviations among tan δ of the consideredCMCs are observed around at $\omega < 2.5$ rad/s while it keeps almost the same when ω exceeds 2.5rad/s. This interesting phenomenon to tan δ shows that the effect of M_{cs} on the frequency-dependence behavior of CMCs is more sensitive at low frequency range.

Effect of Chitosan Molecular Weight on the Steady Rheological Properties of CMCs at Highly Hydrated State

Shear viscosity (η) as a function of the applied shear rate (γ) ispresented in Figure 4A. In our previous work, two unique transitions denoting a sharp drop of the shear viscosity at low shear rate and a small peak of the shear viscosity at high shear rate were observed. The presence of these transitions was closely related to the adsorbed content ofchitosan. [33]Because of the high chitosan/clay weight ratio used here, only the small peak, whichis corresponding to the yield region of the network structure of CMCs, is observed in the presentcase. Interestingly, the center of the small peaks shifts from 5.7 s^1 for CMCs-44 to 12.4 s^{-1} forCMCs-330. This shift in the yield region of CMCs to high shear rate indicates that the network structure formed through high M_{cs}chitosan possesses a stronger ability to resist the external shear flow than those of formed by low M_{cs} one. A further proof on this shift behavior isachievable from Figure 4B, which presents the shear stress as a function of the applied shear rate.Corresponding to that of the shear viscosity, shift in the peak region of the shear stress are also observed. Moreover, the shear stress has a sharp drop following the peak region. There is about 2.5 times drop in the shear stress at = 100 $s\dot{\gamma}^{-1}$ as compared to that at the center of the peakregion. As shown by the inserted Figure c in Figure 4A, the sharp drop of the shear stress is attributed to the disorganization of the network structure at high shear rate. The state of CMCs inthe inserted Figure a and b denotes a partial orientation and deformation of the network at low shear rate and a full deformation of the network at its yield region, respectively.[33] This nonlinear rheological behavior of CMCs is distinguished from that of the clay suspension and chitosan aqueous solution, which usually exhibit a nearly linear shear-thinning behavior and Newtonian behavior, respectively. [43-45]

Effect of Chitosan Molecular Weight on the Thixotropic Properties of CMCs at Highly Hydrated State

Thixotropic behavior observed in clay-water suspension is a time-dependent process associated with restoration and destruction of the structure of clay dispersion systems.This process has been found to closely relate to pH, [46, 47] mechanical history, [48] ionic strength, [49] charge density [50] and clay content [51, 34] in clay-water systems. And the time scale of measurementalso plays important role on the type of CMCs-330 shows a typical linear increase, which is wellmatch a power law stating thixotropic flow of clay suspension. [52, 53] The recovery of thesuspension at rest after the steady shear was usually considered as a basic thixotropic process. To study the effect of M_{cs} on the thixotropic behavior of CMCs, recovery of CMCs after underwenta pre-shearing treatment is tested by the oscillatory shear measurement in the linear viscoelastic region. Changes in G'_{After}, which denotes the storage modulus of CMCs-44, CMCs-130 and CMCs-330 after underwent a pre-shearing process at = 1 $\dot{\gamma}^{-1}$ for 300 s, are depicted in Figure 5as a function of the testing time. Monotonic increase of G'_{After} for all of the considered CMCswithin the time scale of measurement is observed, which indicates the basic thixotropic behavior of CMCs. Interestingly, the recovery behavior of G'_{After} within the testing time is closely related to the molecular weight of chitosan. It implies that the molecular weight of chitosan has the ability to affect the restoring capability of CMCs. G'_{After} of CMCs-330 shows a typical linear increase, which is well match a power law stating:

$$G'_{After} \sim t^n \quad (2)$$

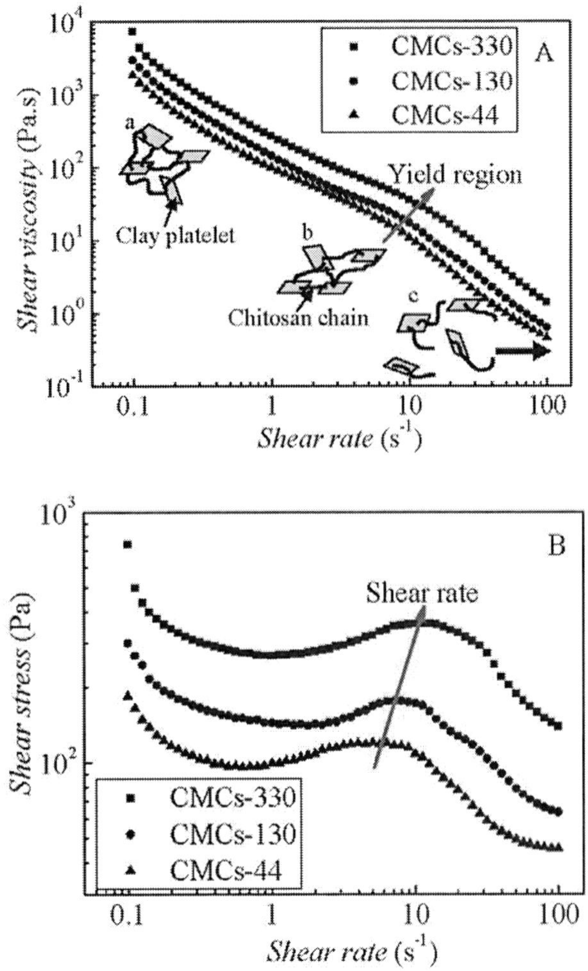

Figure 4. Shear viscosity (η) (A) and shear stress (B) as a function of the shear rate (0.1~100 s^{-1}) at steady-state shear mode for CMCs with different chitosan molecular weight. The inserted Figure (a-c) in Figure 4A shows the different state of CMCs during the steady shearing process, (a) partial orientation and deformation of CMCs at low shear rate, (b) full deformation of the network structure at its yield region, (c) breakage of the network structure at high shear rate.

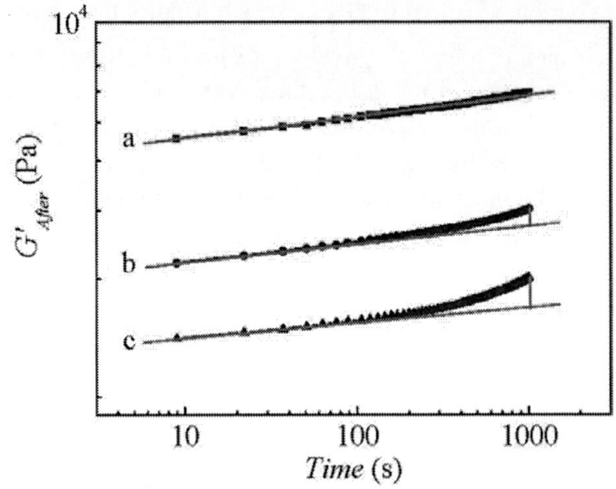

Figure 5. Time dependence of $G'_{After}$for CMCs-330 (a), CMCs-130 (b) and CMCs-44 (c) after pre-sheared at 1 $s^{-1}$for 300s.

This power law is usually applied to characterize the kinetics of the thixotropic recovery of clay suspensions. Within the testing time, the exponent n for CMCs-330 is fitted as 0.078 from theslope of log G'_{After}*vs* log t plot. However, this power law is not acceptable to describe the relationship between log G'After ,and log t for CMCs-130 and CMCs-44 because of the exponentlike increase in their G'_{After}. The exponential increase becomes more distinct at low M_{cs}, which give rise to the remarkable deviation of log G'_{After} vs log t plot from the power law. The rapidly exponential increase in G'_{After} of CMCs-44 suggests that CMCs prepared from low molecular weight chitosan has strong recovery capability in its microstructure. This nonlinear thixotropic behavior of CMCs with low M_{cs} is not in accordance with those linear behaviors observed inclay-water suspension.[35] However, a more detailed discussion on this phenomenon is not possible based on our present data, but it seems that the recovery process of CMCs is related to the length of the adsorbed chitosan chain. No equilibrium is reached within the testing time for all of the samples.

The effect of pre-shearing history on the thixotropic properties of CMCs is further studied.Figure 6A presented are the recovery curves of G'_{After} for CMCs-330 obtained at the preshear rate of = 0.1, 1, 10, 30, 50 and 100 $s{\bullet}\gamma^{-1}$. It is quite clear that G'_{After} of CMCs-330 shows strong dependence on the applied steady shear. There is a transition in the slopes of the recovery curvesfrom positive to negative at the high pre-shearing rate, which indicates the weakening of the recovery capability in the microstructure of CMCs due to the applied strong pre-shearing flow.For a further description, G'_{After} values of CMCs from the end of each recovery test are plotted asa function of the applied pre-shear rates in Figure 6B. The result clarifies that G'_{After} for all of theCMCs decreases firstly with the increase of the applied shear rate and then reaches a plateauwhen the applied pre-shear rate exceeds a certain value. Approximately, the critical values of forCMCs-330, CMCs-130 and CMCs-40 are determined at 43, 30 and 25 $s\dot{\gamma}^{-1}$, respectively. Combined with the result shown in Figure 4A, this result further indicates that the ability ofCMCs to resist the external shear flow is weakened at low M_{cs}. In the plateau regime, full disorganization of the network structure and full orientation of the CMCs particles are expectable under the strong steady shear flow, which would cause more viscous-like of CMCs.

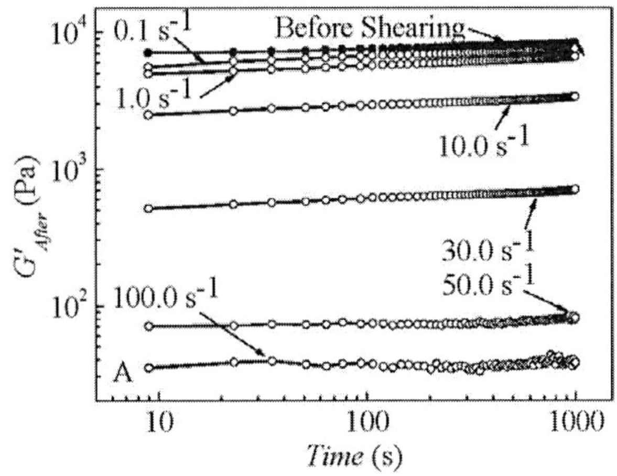

A.

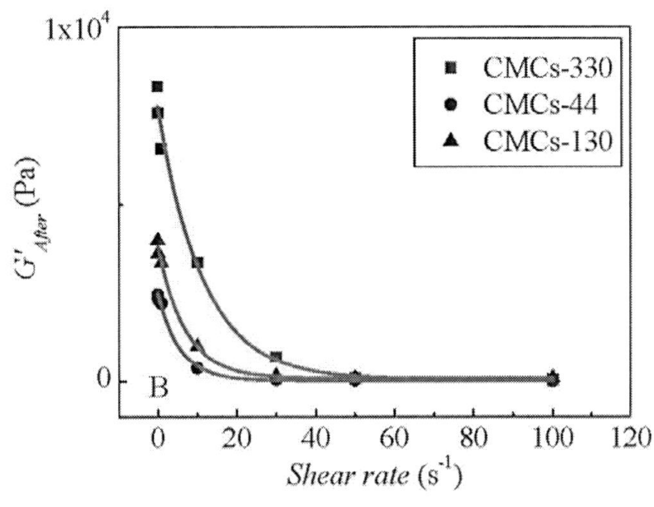

B.

Figure 6. (A) Time dependence of G'_{After} of CMCs-330 after underwent the pre-shearing treatment at 0.1, 1.0, 10, 30, 50 and 100 s^{-1} for 300s; (B) G'_{After} of CMCs as a function of the pre-shearing rate at the end of the recovery time.

Test of Cox-Merz Rule on the Rheological Behavior of CMCs at Highly Hydrated State

Empirical Cox-Merz rule is further applied to analyze the ability of CMCs in alternating theirmicrostructure and in orientating their basic particles under the external shear flow. Concisely, only the analysis results of CMCs-44 and CMCs-330 are demonstrated in Figure 7. The Cox-Merz rule, which is generally acceptable for homopolymers and also some unfilled disordered liquid-like block copolymers, states that

$$\eta^*(\omega) = \eta(\dot{\gamma}) \text{ for } \omega = \dot{\gamma} \qquad (3)$$

where, $\eta^*(\omega)$ and $\eta(\dot{\gamma})$ are the frequency-dependence complex viscosity and the shear viscosity,respectively. However, it is not applicable for the highly filled polymer matrix, in which $\eta^*(\dot{\gamma})$ always higher than $\eta(\dot{\gamma})$. The magnitude of the difference between $\eta^*(\dot{\gamma})$ and $\eta(\dot{\gamma})$. was found toincrease with the increase of silicate loading. [54, 55] Figure 7 shows that $\eta^*(\omega)$ always exceeds $\eta(\dot{\gamma})$ over the entire range of the shear flow for both of CMCs-44 and CMCs-330, which indicates a failure to the Cox-Merz rule of our system. According to those highly filled polymer systems examined previously, the failure for CMCs here likely stems from some preferential orientationof the initial quiescent random arrangement of CMCs particles or their tactoids under a steadyshear. To further demonstrate the orientation of CMCs under the shear flow, the decreaseamplitude (Δη) between η^* and $\eta(\dot{\gamma})$ is calculated as follow

$$\Delta\eta = [\eta(\omega) - \eta(\dot{\gamma})] / \eta(\omega) \text{x } 100\% \qquad (4)$$

The calculated results indicate that, with the increase of the applied shear rate, Δη increases from 82.5% and 91% to 98% and 99% for CMCs-44 and CMCs-330, respectively. This increase in Δη suggests the orientation of CMCs is enhanced at high shear rate. Moreover, Δη of CMCs330 is smaller than that of CMCs-44 at low shear rate, indicative of a weaker orientatingcapability of CMCs-330. However, at high shear rate, both of CMCs-44 and CMCs-330 show a quite similar Δη, suggesting a similar orientating capability in their microstructure. High values of Δη also further indicate a remarkable breakage and orientation in the network microstructure and particles of CMCs as a result of the applied strong shear flow.

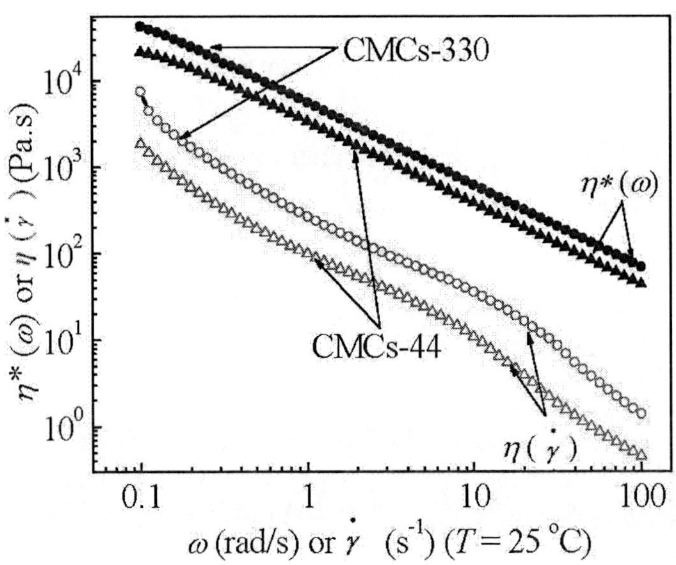

Figure 7. Test of Cox-Merz rule on the rheological behavior of CMCs-330 and CMCs-44.

CONCLUSION

The effect of M_{cs} on the rheological behavior of CMCs at highly hydrated state has been studied. The critical strain referring to the transition from elastic-like to viscous-like of CMCs,was observed around at $\gamma_c = 42$ % for CMCs-44 and shifted to about 54% for CMCs-330. It indicated that the capability of CMCs to resist the external strain was enhanced at high M_{cs}. Thevalue of tan δ ranging from 0.1 to 0.5 over the entire range of the applied frequency suggested adominant elastic response of CMCs and a strong associative interaction in CMCs. The slopetransition from negative to positive of the tan δ vs log ω plot at ω = 2.5 rad/s implied a quick increase of viscous-like structure in CMCs at high frequency. Moreover, all of the CMCs exhibited typical shear thinning behavior. Small peak region in the shear viscosity, which is corresponding to the yield region of the CMCs network, shifted to high shear rate with the increase of M_{cs}. It further confirmed the enhanced mechanical stability of the network formed athigh M_{cs}. Furthermore, thixotropic behavior was clarified to closely depend on M_{cs}. With n = 0.078, G'_{After} of CMCs-330 showed a linear and monotonic increase within the testing time, which was well fitted to the power law of $G'_{After} \sim t^n$. However, this power law was unacceptableto that of CMCs-130 and CMCs-44 as a result of their exponent-like increase. The deviation ofG'_{After} from the power law became more remarkable with the decrease of M_{cs}, indicating arelatively strong recovery capability of CMCs at low M_{cs}. Moreover, pre-shearing history test indicated that the recovery ability of CMCs could be weakened by increasing the applied pre-shearing rate. Empirical Cox-Merz rule was not applicable for all of the CMCs. It therefore clarified that the shear-thinning behavior of CMCs was due to some preferential orientation ofthe initial quiescent random arrangement of CMCs particles or their tactoids under the shear flow. Orientation of CMCs-330 was less sensitive to the low shear rate as compared to that ofCMCs at low M_{cs}.

ACKNOWLEDGMENTS

We thank Dr. Jozef Kokini for the use of rheology instruments. This work was supported by National Research Initiative Grant 2006-35503-17568 from the USDA Cooperative State Research, Education, and Extension Service Program on Improving Food Quality and Value.

REFERENCES

[1] Giannelis, E. P. *Adv. Mater.* 1996, 8, 29-35.

[2] Suter, J. L.; Coveney, P. V.; Greenwell, H. C.; Thyveetil, M.-A. *J. Phys. Chem.* C 2007,111, 8248-8259.

[3] Messersmith, P. B.; Giannelis, E. P. *Chem. Mater.* 1993, 5, 1064.

[4] Okamotoa, M.; Moritaa, S.; Taguchia, H.; Kima, Y. H.; Kotakaa, T.; Tateyamab, *H. Polymer* 2000, 41, 3887-3890.

[5] Chasteka, T. T.; Steina, A.; Macosko, C. Polymer 2005, 46, 4431-4439.

[6] Vaia, R. A.; Ishii, H.; Giannelis, E. P. *Chem. Mater.* 1993, 5, 1694.

[7] Ogata, N.; Jimenez, G.; Kawai, H.; Ogihara, T. J. Polym. Sci. Part B: *Polym. Phys.* 1997,35,389.

[8] Pavlidou, S.; Papaspyrides, C. D. *Progress in Polymer Science* 2008, 33, 1119-1198;
[9] Geoffrey M. Bowers, David L. Bish, and R. James Kirkpatrick, *J. Phys. Chem. C*2008,112, 6430-6438.
[10] Kornmann, X.; Lindberg, H.; Berglund, L. A. P*olymer* 2001, 42, 1303;
[11] Manias, E.; Touny, A.; Wu, L.; Strawhecker, K.; Lu, B.; Chung, T. C. *Chem. Mater.2001*,13,3516.
[12] Jacobs, J. D.; Koerner, H.; Heinz, H.; Farmer, B. L.; Mirau, P.; Garrett, P. H.; Vaia, R. A. *J. Phys. Chem. B* 2006, 110, 20143-20157.
[13] Vaia, R. A.; Giannelis, E. P. *Macromolecules* 1997, 30, 8000.
[14] Zerda, A. S.; Lesser, A. J. *J. Polym. Sci. Part B:Polym. Phys.* 2001, 39, 1137.
[15] Kim, C. M.; Lee, D. H.; Hoffmann, B.; Kressler, J.; Stoppelmann, G. *Polymer*2001,42,1095.
[16] Huang, W.; Han, C. D. *Macromolecules* 2006, 39, 257-267.
[17] Lee, K.; Han, C. D. *Macromolecules* 2003, 36, 8796-8810.
[18] Krishnamoorti, R.; Giannelis, E. P. *Macromolecules* 1997, 30, 4097-4102.
[19] Lagaly, G. *Phil. Trans. R. Soc. Lond. A* 1984, 311, 315-332.
[20] Bujdák, J.; Komadel, P. *J. Phys. Chem. B,* 1997, 101, 44, 9065-9068.
[21] Lee, S. Y.; Echo, W. J.; Kim, K. J.; Ahn, J. H.; Lee, M. J. *Colloid Interface Sci.* 2005,284,667-673.
[22] Park, S. J.; Seo, D. I.; Lee, J. R. *J. Colloid Interface Sci.* 2002, 251, 160-165.
[23] Sasai, R.; Itoh, H.; Shindachi, I.; Shichi, T.; Takagi, K. *Chem. Mater.* 2001, 13, 2012-2016.
[24] Chivrac, F.; Pollet, E.; Schmutz, M.; Avérous L. *Biomacromolecules* 2008, 9, 896-900.
[25] Park, H.; Liang, X.; Mohanty, A. K.; Misra, M.; Drzal, L. T. *Macromolecules* 2004,37,9076-9082.
[26] Park, H.; Misra, M.; Drzal, L. T.; Mohanty, A. K. *Biomacromolecules* 2004, 5, 2281-2288.
[27] Günister, E.; Pestreli, D.; Ünlü, C. H.; Atıcı, O.; Güngör, N. *Carbohydrate Polymers*2007,67,358-365.
[28] Darder, M.; López-Blanco, Aranda, M. P.; Aznar, A. J.; Bravo, J.; Ruiz-Hitzky, E. *Chem. Mater.* 2006, 18, 1602-1610
[29] Wang, S.; Chen, L.; Tong, Y, J. *J. Polym. Sci. Part A: Polym. Chem.* 2006, 44, 686.
[30] Xu, Y.; Ren, X.; Hanna, M. A. J. *Appl. Polym. Sci.* 2006, 99, 1684-1691.
[31] Darder, M.; Colilla, M.; Ruiz-Hitzky, E. *Chem. Mater.* 2003, 15, 3774.
[32] Tang, C.; Xiang, L.; Su, J.; Wang, K.; Yang, C.; Zhang, Q.; Fu, Q. *J. Phys. Chem.* B2008,112, 3876-3881.
[33] Liang, S.; Liu, L.; Huang, Q.; Yam, K. L.*; J. Phys. Chem. B Accepted.*
[34] Pignon, F.; Magnin, A.; Piau, *J. J. Rheol.* 1998, 42, 1349-1373.
[35] Li, S.; Hou, W.; Sun, D.; Guo, P.; Jia, C. *Langmuir 2003*, 19, 3172-3177.
[36] Agrawal, S. K. Sanabria-DeLong, N.; Tew, G. N.; Bhatia, S. R. *Langmuir,* 2008, 24, 13148-13154.
[37] Haraguchi, K; Takehisa, T. *Adv. Mater.,* 2002, 14, 1120;
[38] Haraguchi, K.; Takehisa, T.; Fan, S. *Macromolecules,* 2002, 35, 10162.
[39] Krishnamoorti, R.; Giannelis, E. P. *Langmuir* 2001, 17, 1448-1452.
[40] Gelfer, M. Y.; Burger, C.; Chu, B.; Hsiao, B. S.; Drozdov, A. D.; Si, M.; Rafailovich, M.;Sauer, B. B.; Gilman, J. W. *Macromolecules* 2005, 38, 3765-3775.
[41] Winter, H. H.; Mours, M. *Adv. Polym. Sci.* 1997, 134, 166-233.

[42] Rohn, C. L. Analytical Polymer Rheology*; Structure Processing Property Relationships*; Hanser Publisher: Munich, 1995, 202-204.
[43] Shalkevich, A.; Stradner, A.; Bhat, S. K.; Muller, F.; Schurtenberger, P. *Langmuir,* 2007, 23, 3570-3580.
[44] Baravian, C.; Vantelon, D.; Thomas, F. *Langmuir 2003*, 19, 8109-8114.
[45] Desbrieres, J. *Polymer* 2004, 45, 3285-3295.
[46] Wells, N.; Childs, C.W. *J. Soil Res.* 1988, 26, 145-152.
[47] Thomson, D. W.; Butterworth, J. T. *J. Colloid Interface Sci.* 1992, 151, 236.
[48] Joye, D. D.; Poehlein, G. W. *Transactions of the Society of Rheology* 1971, 15, 51-61.
[49] Mourchid, A.; Delville, A.; Lambard, J.; Lécolier, E.; Levitz, P. *Langmuir* 1995, 11,1942-1950.
[50] Christidis, G. E.; Blum, A. E.; Eberl, D.D. *Applied Clay Science* 2006, 34,125-138.
[51] Pignon, F.; Magnin, A.; Piau, J. *J. Rheol.* 1996, 40, 573-587.
[52] Nakaishi, K.; Yasutomi, R. *Applied Clay Science* 1994, 9, 71-79.
[53] Pignon, F.; Magnin, A.; Piau, J.; Cabane, B.; Lindner, P.; Diat, O. *Phys. Rev. E* 1997,56,3281-3289.
[54] Larson, R. *The Structure and Rheology of Complex Fluids;* Oxford University Press: NewYork, 1999.
[55] Ren, J.; Krishnamoorti, R. *Macromolecules* 2003, 36, 4443-4451.

In: Chemical and Biochemical Kinetics: New Perspectives ISBN 978-1-61209-224-9
Editor: Gennady Zaikov

Chapter 20

MECHANISM OF STABLE RADICAL GENERATION IN LIGNIN UNDER THE ACTION OF NITROGEN DIOXIDE

E. Ya. Davydov,* I. S. Gaponova, S. M. Lomakin, G. B. Pariiskii, T. V. Pokholok, and G. E. Zaikov

N. M Emanuel Institute of Biochemical Physics of Russian Academy of Sciences, Moscow, Russia

ABSTRACT

Lignin is the very sensitive to the nitrogen dioxide actionas evidenced by reasonably high concentrations of stable radicals which can be accumulated in exposed samples (~2 x 1019 spins/g). The generation of nitrogen containing radicals is connected with oxidative electron-transfer reactions initiated by dimers of NO_2 in phenol units and nonphenolic structures. The subsequent radical conversions include processes of degradation and modification (nitrosation and nitration) of a lignin.The formation of stable iminoxyl radicals in the presence of NO_2 is the typical process for lignin.

Keywords: nitrogen dioxide, lignin, stable radicals, kinetics, mechanism, application

INTRODUCTION

N.M. Emanuel Institute of Biochemical Physics (Russian Academy of Sciences) has several perfect installation of electron spin resonance (ESR). ESR-method is very effective and very sensitive for investigation of radical reaction in chemistry and biochemistry. This method was created in 1944 [1] by Prof. K.M. Zavoiskii (Academy of Sciences of USSR). V.V. Voevodskii, L.A. Blumenfeld, Ya.S. Lebedev, E.L. Frankevich, A.L. Buchavhenko,

* E-mail: Chembio@sky.chph.ras.ru.

E.G. Rozantsev, A.M. Wasserman and A.L. Kovarskii did very much for application of this method for investigation of chemical and biochemical reactions [2-8].

Aim and Scope

The contributors of this paper published in 2009 the book [4] about interaction of polymers with polluted atmosphere (nitrogen oxides). Our new paper is devoted for the problems of interaction between nitrogen dioxide and lignin by ESR. Mankind accumulated tremendous amount of lignin and did not find real ways for applications of it.

So, any kind of research about lignin (applications) is very important for looking for the fields of utilization.

Methods of Experiments

All information about application of ESR as well as spectroscopy for investigation of reaction between nitrogen dioxide was published recently [3].

Experimental Results and Discussion

Detection of specific stable radicals by ESR in plants can be considered as a sensitive method of air pollution monitoring. The stable radicals are associated with products of free radical reactions initiated by various air pollutants, in particular nitrogen oxides.

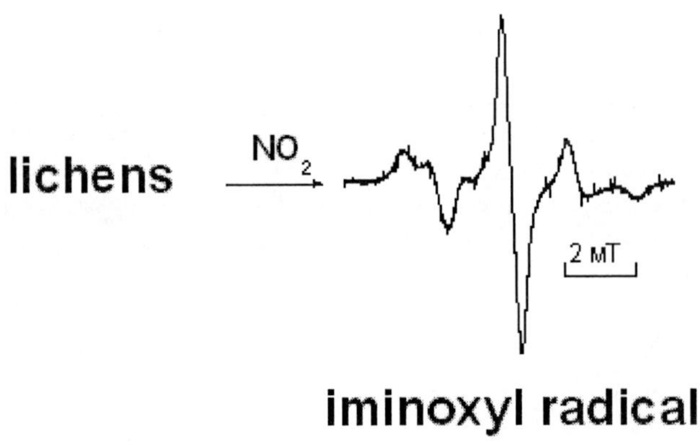

The nitrogen dioxide pre-treatment of wood pulp before oxygen bleaching leads to appreciably decreased lignin contents after a given duration of the oxygen bleaching (Figure 1).

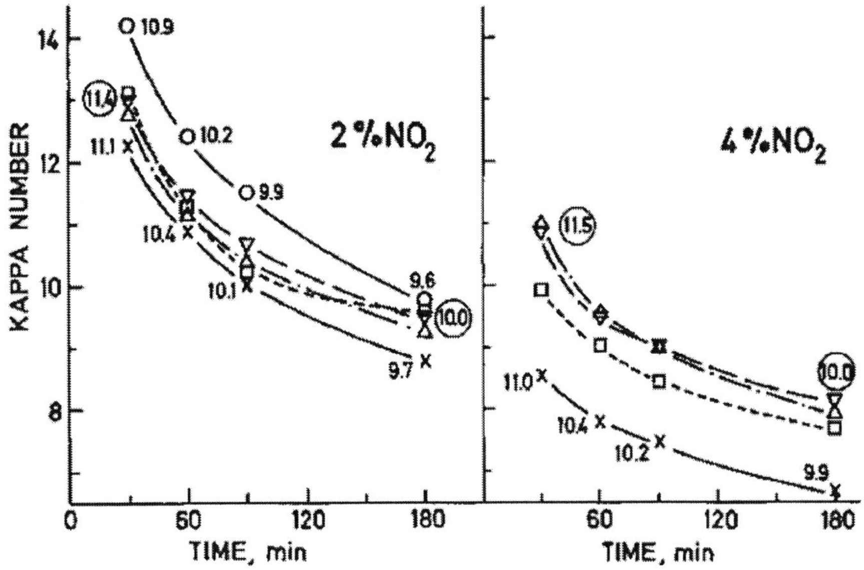

Figure 1. Dependence of kappa number against time for different conditions (2% $NO_2$and 4% NO_2).

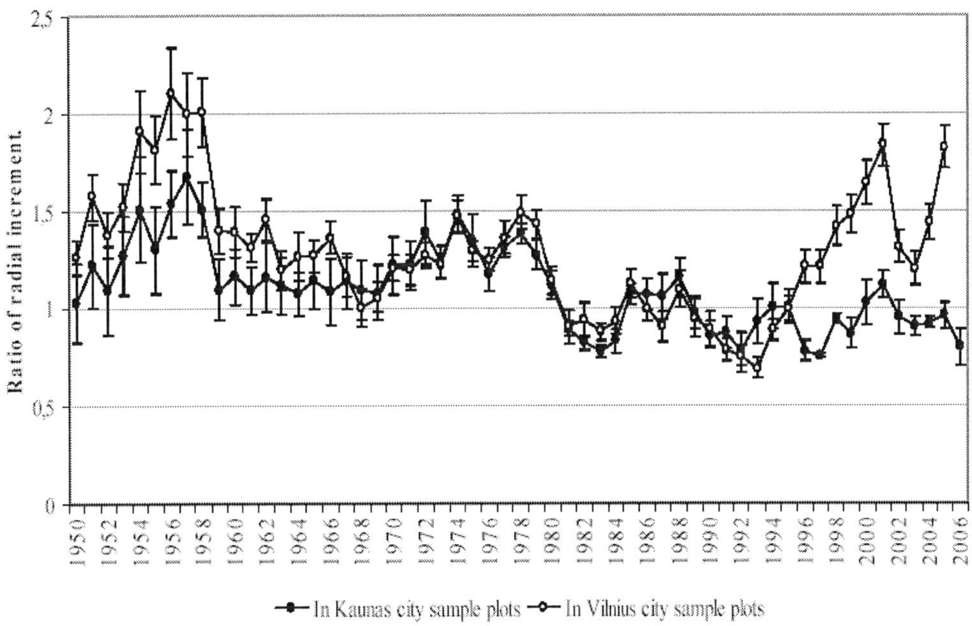

Figure 2. Dependence of ratio of radical increment on the year in Kaunas and Vilnius cities.

The impact of NO_2 on pines growing in cities (Vilnius, Kaunas) has been revealed. The reduction in NO_2 concentration in the atmosphere (1990 – 2006) determines an increase in pine radial increment. It is quite possible that this result is conditioned by interaction of NO_2 with reactive groups of lignin (Figure 2).

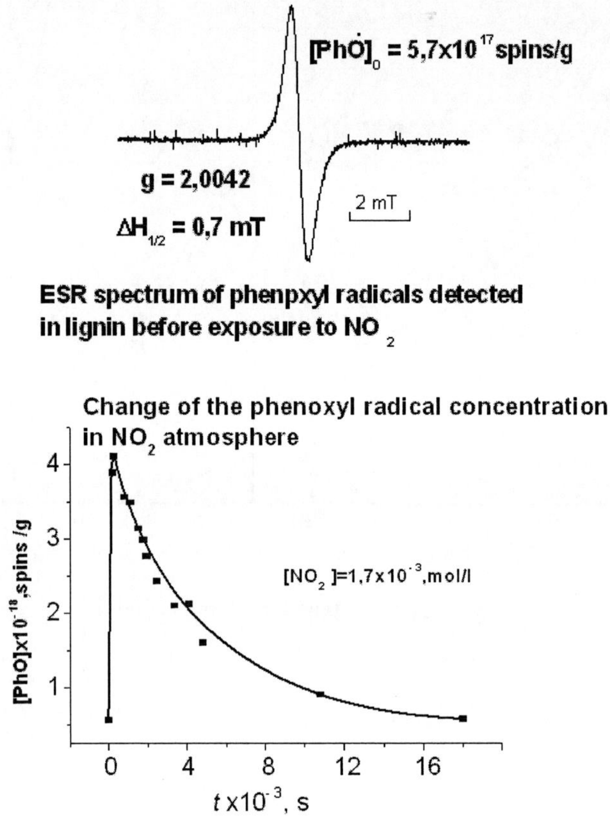

Figure 3. Dependence of phenoxyl radical concentration on time in nitrogen dioxide atmosphere.

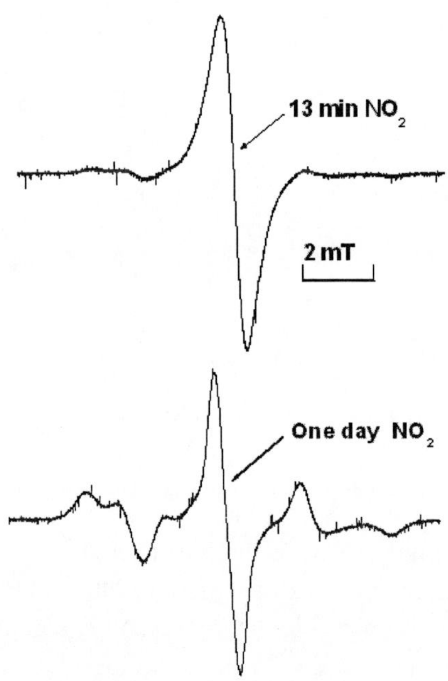

Lignin is evidently reactive to nitrogen dioxide and capable of destroying in NO2 atmosphere. The purpose of the present research was to examine the mechanism of primary radical reactions determining subsequent conversions of lignin units (Figure 3).

Decay of phenoxyl radicals is accompanied by appearance of iminoxyl radicals (Figure 4).

Change in the iminoxyl radical concentration in the course of thermolysis of lignin exposed to NO_2 at room temperature. Figure 5 gives information about concentration of iminoxyl radicals in different temperatures.

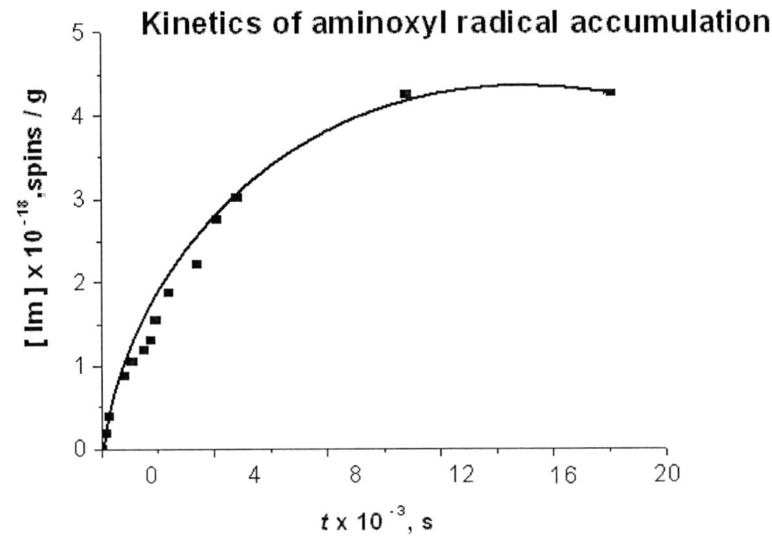

Figure 4. Accumulation of aminoxyl radicals in time.

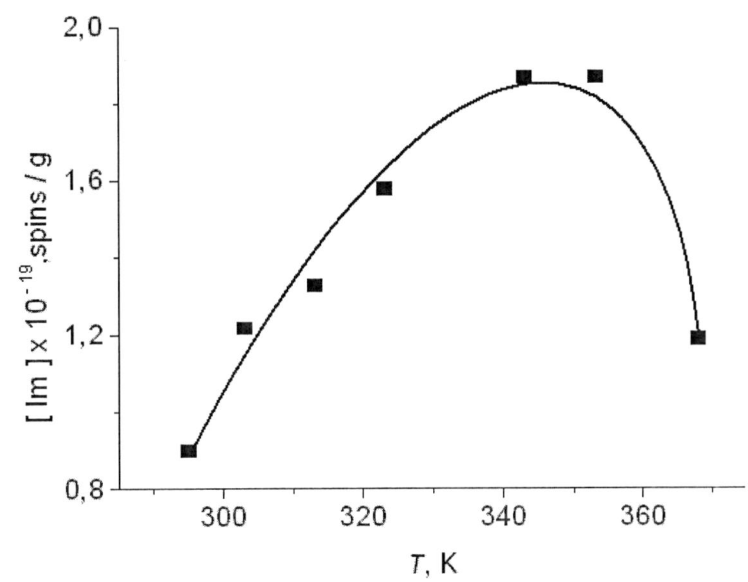

Figure 5. Concentration of iminoxyl radicals for different temperatures.

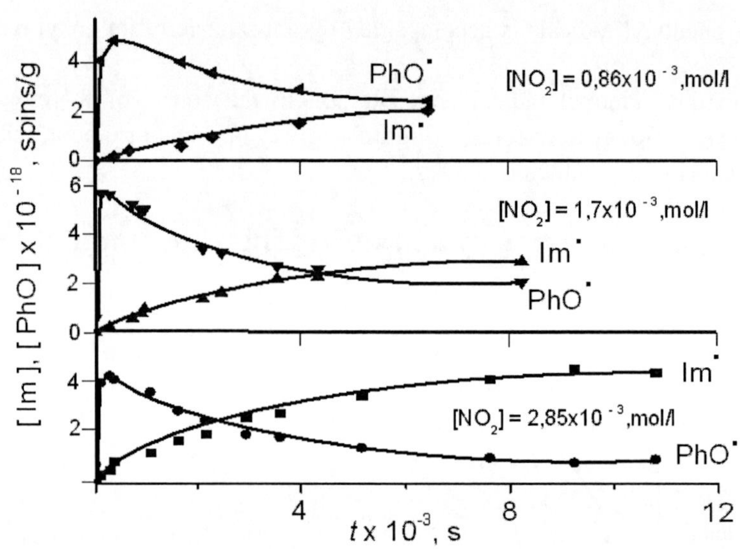

Figure 6. Dependence iminoxyl radical accumulation and phenoxyl radical decay in time.

Thermolysis shifts the equilibrium decay of iminoxyl radicals by recombination with NO_2 to stable radicals.

$$\dot{Im} + NO_2 \rightleftharpoons ImNO_2$$

Rates of iminoxyl radical accumulation and phenoxyl radical decay are approximately equal at various NO_2 concentrations (Figure 6).

Mechanism of nitrosation of monomethoxyphenol groups of lignin with the formation of iminoxyl radicals. Initiators are dimers of NO_2 in the form of nitrosyl nitrate.

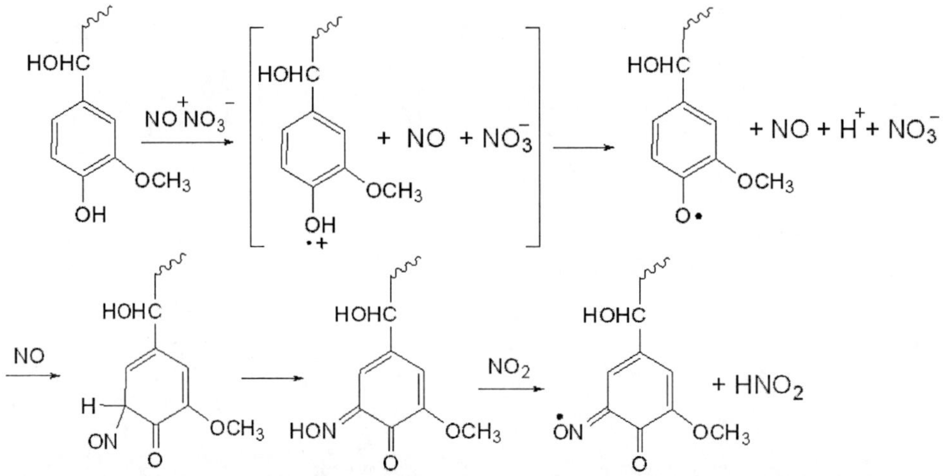

In parallel with conversions of phenols, nitrosyl nitrate is capable of oxidising multitude of hydroxyl groups of nonphenolic structures linking aryl rings in lignin. As a result, aldehydes are formed.

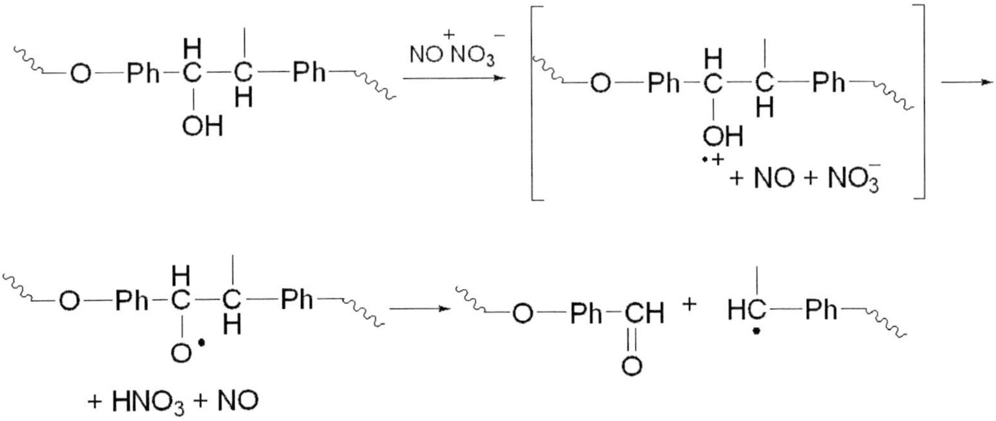

Changes of IR spectra on exposure of lignin to NO_2 confirm the presented mechanism (Figure 7).

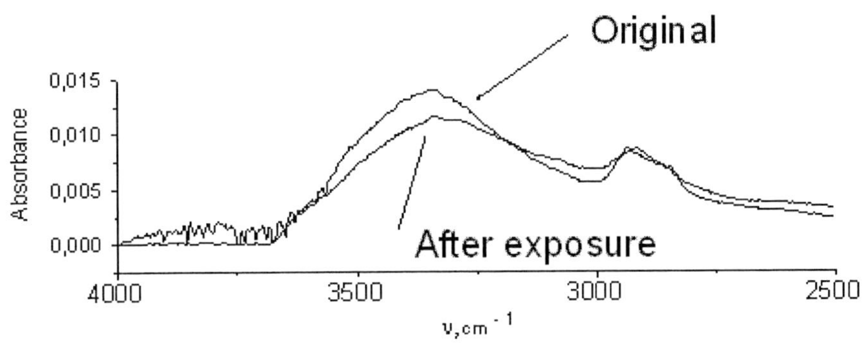

Figure 7. Infrared spectra of lignin before and after interaction with nitrogen oxides.

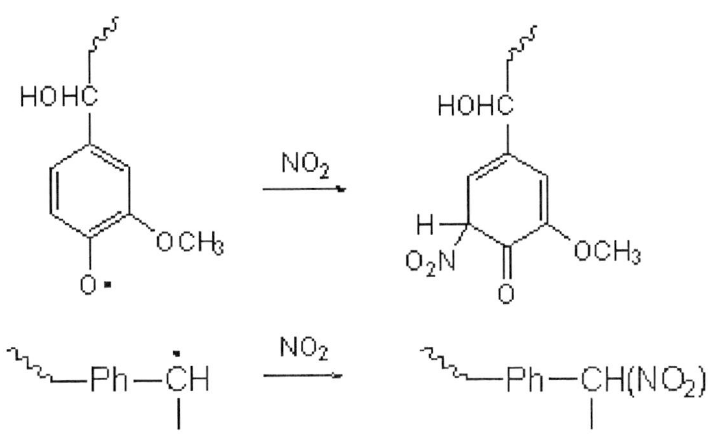

Aldehyde groups should be accumulated in the course of the lignin exposure to NO_2. The band of hydroxyl groups in lignin (3300cm−1) appreciably reduces for two days of NO_2 exposure. In addition, the decrease in intensity is observed for bands corresponding to stretching vibrations of C=C bonds of phenyl rings (1512 and 1450cm−1). Moreover, the appearance of new intense bands at 1558 and 1337cm−1 belonging to asymmetric and symmetric stretching vibrations of N=O bonds in nitro groups takes place.

Aldehydes are precursors of acylaminoxyl radicals observed after prolonged exposure of lignin to NO_2.

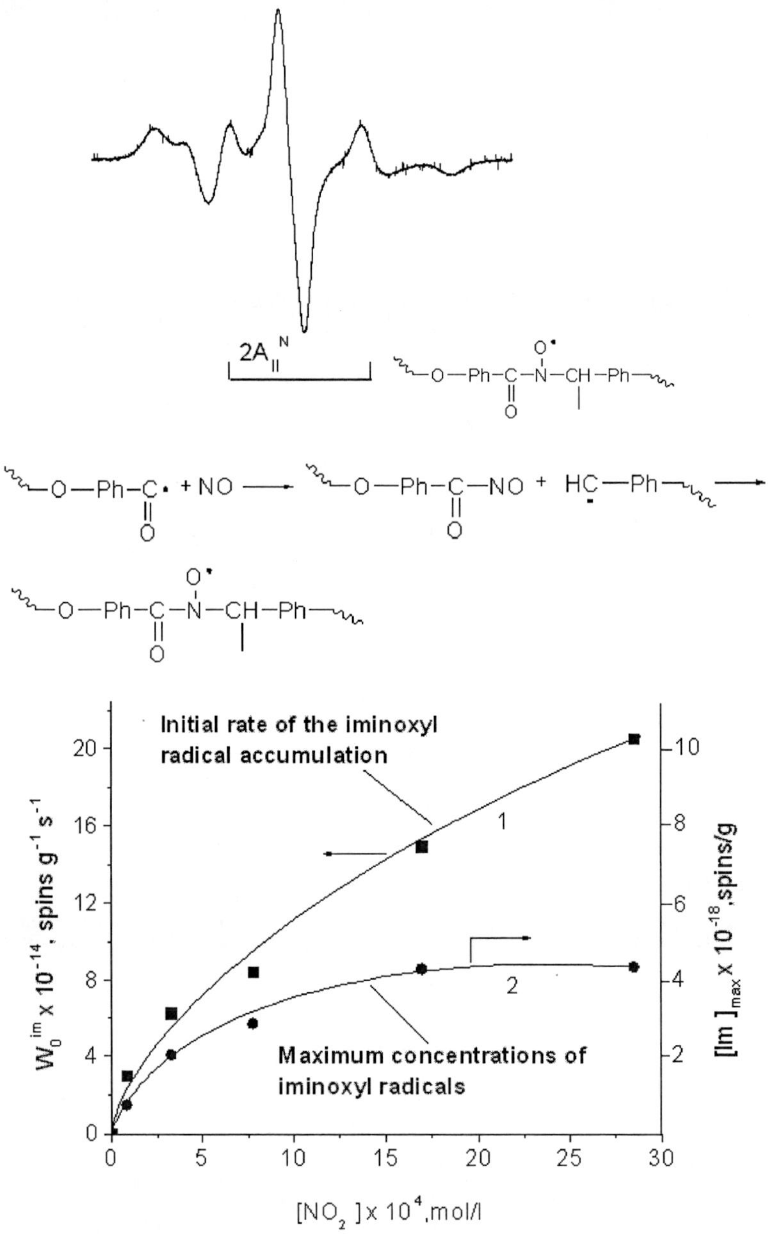

Figure 8. Dependence of accumulation rate of iminoxyl radicals on concentration of nitrogen dioxide.

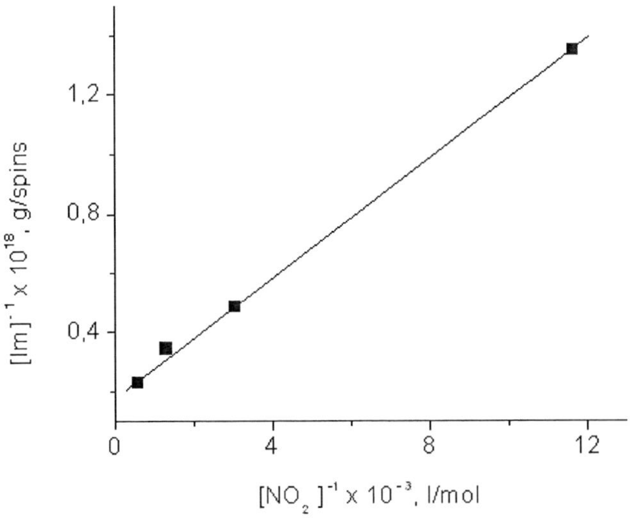

Figure 9. Dependence of the iminoxyl radical yield in lignin on concentration of nitrogen dioxide.

Dependencies of the iminoxyl radical yield in lignin on NO_2 concentrations is presented on the figures 8 and 9.

$$[R_{im}]_{st} = \frac{[R_{im}]_{st}^{max} \cdot b \cdot [NO_2]}{1 + b \cdot [NO_2]}$$

If to plot concentrations of Im • on steady-state levels ([Im •]st) on NO_2 concentrations, the dependence reminding in form the Langmuir isotherm can be obtained. T the adsorption of nitrogen dioxide on the lignin surface is the rate-determining factor for the iminoxyl radical generation.

CONCLUSION

Lignin is the very sensitive to the nitrogen dioxide actionas evidenced by reasonably high concentrations of stable radicals which can be accumulated in exposed samples (~2 x 1019 spins/g).

The generation of nitrogen containing radicals is connected with oxidative electron-transfer reactions initiated by dimers of NO_2 in phenol units and nonphenolic structures. The subsequent radical conversions include processes of degradation and modification (nitrosation and nitration) of a lignin.

The formation of stable iminoxyl radicals in the presence of NO_2 is the typical process for lignin.

REFERENCES

[1] N.M. Emanuel, G.E. Zaikov, V.A. Kritsman *"Chemical kinetics and chain reactions. Historical aspects",* Nova Science Publishers, New York, 1992, 625 pp.

[2] "Chemical kinetics", Ed. by E.B. Burlakova, A.E. Shilov, S.D. Varfolomeev, G.E. Zaikov, *VSP International Science Publishers, Leiden-Utrecht (The Netherlands),* 2005, 682 pp.

[3] "Biological kinetics", Ed. by E.B. Burlakova, A.E. Shilov, S.D. Varfolomeev, G.E. Zaikov, *VSP International Science Publishers, Leiden-Utrecht (The Netherlands),* 2005, 356 pp.

[4] G.E. Zaikov, E.Ya. Davydov, G.B. Pariiskii, I.S. Gaponova, T.V. Pokholok *"Interaction of polymers with polluted atmospheres. Nitrogen oxides",* Smithers, Shawbury, Shrewsbury, Shropshire, UK, 2009, 270 pp.

[5] G.E. Zaikov "Chemical and biochemical physics. New frontiers", *Nova Science Publishers,* New York, 2006, 272 pp.

[6] "Resent advances in polymer nanocomposites: synthesis and classification", Ed. by S. Thomas, G.E. Zaikov, S.V. Valsaraj, A.P. Meera, *Brill Academic Publishers,* Leiden-Boston (The Netherlands - USA), 2010, 436 pp.

[7] "Resent advances in polymer nanocomposites", Ed. by Thomas, G.E. Zaikov, S.V. Valsaraj, *VSP International Science Publishers,* Leiden- Boston (The Netherlands - USA), 2009, 528 pp.

[8] G.E. Zaikov, S.K. Rakovsky *"Ozonation of organic and polymeric compounds",* Smithers, Shawbury, Shrewsbury, Shropshire, UK, 2009, 414 pp.

In: Chemical and Biochemical Kinetics: New Perspectives ISBN 978-1-61209-224-9
Editor: Gennady Zaikov

Chapter 21

THERMOPHYSICAL PROPERTIES OF ORGANOPLASTICS BASED ON PA-6

A. I. Burya,[1] E. V. Tkachenko[2] and O. P. Chigvintseva[1]
[1]Dniepropetrovsk State Agrarian University, Dniepropetrovsk, Ukraine
[2]Sevastopol National University of Nuclear Energy and Industry, Sevastopol, Ukraine

ABSTRACT

The influence of Arimide-T grade polyimide fibre on thermophysical properties (specific heat capacity, coefficients of heat and temperature conductivity) of organoplastics based on PA-6 has been studied. It has been stated that introducing the organic fibre (OF) into the content of polyamide binder decreases the heat capacity by 1.3-1.5 times, as well as increases the heat conductivity and temperature conductivity of organoplastics (OP) by 7 and 11-37 % correspondently.

Keywords: polyamide, organoplastics, thermophysical properties

Organic thermally resistant fibres with the improved functional properties regarded as the reinforcing fibres of the third generation have recently become extremely significant. Despite being produced in the amount of 1% from all chemical fibres production worldwide, the financial results of their marketing make about 1.1 billion dollars a year [1]. Among such polymer fibres there is a thermo- and fire-resistant Arimide-Tgradepolyimide fibre.

This fibre is characterized by its incombustibility and an exclusive resistance to the effect of high temperatures (up to 300-350°C) during a long period of time without any visible decrease of its properties. Arimide-Tis the only fibre worldwide keeping its parameters in the conditions of radiation and ultraviolet irradiation. An extremely interesting property of this OF which is not observed in other synthetic fibres is almost the complete absence of residual deformations at the increased temperatures. Applying this OF as a reinforcing filler of polymer binders allows one to obtain the goods characterized by their absolute incombustibility and high stability of the dimensions within the broad temperature and moisture range [2].

The investigations that we had held earlier showed that applying Arimide-Tas a reinforcing filler of aliphatic PA-6 allows to obtain OP possessing the improved thermal, physico-mechanical and tribological properties [3-4].

Asitisknown, intheprocessofdefiningthepracticalvalueofthepolymercomposites (PC) a significant part is played by theinvestigationoftheirthermophysicalproperties. A lot of the details of technical devices made of PC are exploited in nonstationary heat fields, that it why investigating their thermophysical properties is a topical task. The processes of heat transfer have a great practical value in aviation, space exploration, industrial energetics, technological processes of chemical, building and other fields of industry [5].

Concerning the above-mentioned, aimed at studying the behaviour of OP based on PA-6 in the conditions of the increased temperatures, the influence of Arimide-Tfibre content on basic thermophysical properties was investigated.

CompositionsfromOPbasedonPA-6 reinforcedwith 15, 30 and 45 mass % ofArimide-Twereobtainedbymeansofdrymixingwithintherotatingelectromagneticfieldwiththeusageofferro magneticparticles, afterwhichtheywereworkedintothe blockware by means of compression moulding. Thespecificheatcapacity(C_p, kJ/kg· K) and the coefficient of heatconductivity (λ, W /m· K) OPwere defined at IT-C-400 andIT-λ-400 measuring units accordingtoGOST 23630.1-79 and 25630.2-79. The coefficient of temperatureconductivity (a, m^2/sec) was defined by the formula: $a = \lambda / (C_p \cdot \rho)$, where$\rho$is the specimen's density, kg/m^3.

The temperaturedependency research of PA-6 and OP based on it (table 1) showed that it linearly grows within quite a wide temperaturerange (173-473 K) depending on the temperature increase (by 3.7 times on average) due to the polymer macromolecules' flexibility increase; at the same time, the heat capacity suddenly changed (ΔC_p) in the area of materials transfer from the glassy state into the high-elasticitystate (448-498 K).

It is known [6] that the heat capacity sudden change valueΔC_p within the area of phase transitions is a semiquantitative measure of boundary layers' content within the system "polymer binder – fibrous filler". In our case, the increase of Arimide-Tfibre content up to 30 mass % resulted in the decrease of theΔC_p value (table 2) compared to the binder, after which it started increasing again as the filler content made up to45 mass %.

Such behaviour of ΔC_p can be explained by the development of two competing processes within the OP structure. The first one lies in the fact that ΔC_p decrease during vitrification in case of Arimide-T concentration growth is connected with the increase of the polymer share involved in transition layers, and with excluding this share of the molecules segments from those participating in vitrification. The drawn conclusion is confirmed by the calculation data referring to the share of macromolecules having transferred into the boundary layers (table 2). The calculation was performed according to the ratio [7]:

$$\left(\frac{\Delta r + r}{r} - 1 \right)^3 = \nu \cdot \frac{\Phi}{1 - \Phi}$$

where: $\nu = 1 - \Delta C_{p,н.} / \Delta C_p$is the share of macromolecules having transferred into the boundary layers;$\Delta C_{p.н.}$ is a sudden change of the filled polymer's specific heatcapacity; ΔC_p is a sudden change of the non-filled polymer's specific heatcapacity; Φ is the bulk content of the filler;$r$is the fibre radius, $\Delta r$is the thickness of the boundary layer, mcm.

Table 1. The influence of temperature on the heat capacity of PA-6 and organoplastics

Material	Temperature, K																
	173	198	226	248	273	298	323	348	373	398	423	448	473	498	523	548	573
PA-6	0,82	1,20	1,48	1,60	1,62	1,64	1,92	1,93	2,17	2,31	2,67	3,01	3,24	2,13	1,43	1,03	0,74
CA15	0,76	1,20	1,35	1,38	1,42	1,59	1,73	1,91	2,10	2,17	2,51	2,61	2,21	1,83	1,34	1,40	1,42
CA30	0,70	1,14	1,23	1,29	1,36	1,57	1,68	1,76	1,91	2,13	2,26	2,40	2,66	1,81	1,29	0,89	0,70
CA45	0,63	0,91	1,10	1,10	1,26	1,53	1,58	1,72	1,91	2,08	2,23	2,38	2,53	1,66	1,10	0,66	0,52

Note: CA15, CA30 and CA45 – PA-6 reinforced with Arimide-T fibre at the amount of 15, 30 and 45 mass % correspondently.

Table 2. Calculatedthermophysicalpropertiesof organoplastics

Properties	Content of Arimide-T fibre, mass %		
	15	30	45
Sudden change of specific heat capacity, kJ/kg · K	0.76	0.59	0.72
Share of polymer macromolecules having transferred into the boundary layers	0.03	0.24	0.08
Thickness of the boundary layer, mcm	0.34	1.75	0.60

Note. A sudden change of PA-6 specific heat capacity has made 0.78 kJ/kg · K.

The second process occurring within the OP structure lies in the fact that Arimide-T fibre content increase leads to the loosening of polymer molecular packing within the boundary layers, which results in decreasing the number of contacts between the fragments of macromolecules chains accompanied with the mobility growth of the latters. As a result of this, the sudden change of heat capacity increases, and the share of polymer macromolecules having transferred into the boundary layers decreases. The course of these processes has a correspondent influence on the thickness of the boundary layer: at OF content up to 30 mass % it grows from 0.34 to 1.75, after which its dropping to 0.6 mcm can be observed (table 2).

In general, the increase of OF amount within PC allows one to decrease the specific heat capacity by 1.3-1.5 times. Encrathy calculation of C_p/T – the temperature coefficient of enthropy which most significantly reflects the changes of the macromolecule structural elements' mobility [8] – showed that in case of introducing OF into the polyamide matrix within the investigated temperaturerange this index is decreased by 30-33% (Figure 1a).

The temperature dependency of OP heatconductivitycoefficient (Figure 1б) being a thermophysical parameter connected with the distribution and dispersion of elastic waves caused by heat fluctuations of the body particles testifies to its growth with the increase of temperature. Inthiscase, a weak maximum of temperature conductivity is observed within the area of OP transition from the glassy state into the high-elasticity state (448-498 K). The concentration dependency of PC change sambiguously: in the temperature range of 173-323 K, at Arimide-T fibre content growth, the heat conductivity of OP increases on average by 7 %, at 348-423 K is has little influence on filler (FL) content and gets the values of 0.38-0.40 W/m· K, while in the temperature range of 448-498 K the maximum values of λ are characteristic of OP containing 15 mass % of OF.

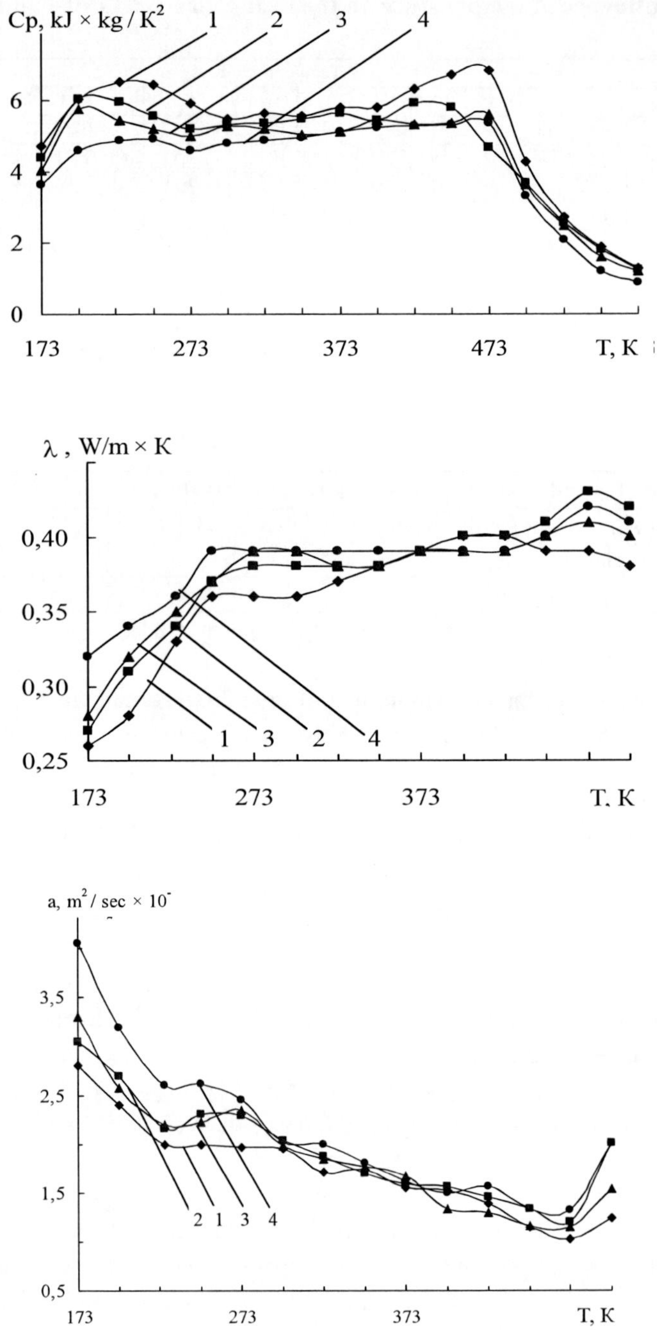

a.

b.

c.

Figure 1. Temperature dependency of encrathy (a), coefficients of heat conductivity (b) and temperature conductivity (c) of PA-6 (1) and organoplastics based on it reinforced with 15 (2), 30 (3) and 45 (4) mass % of Arimide-TThe latter can be explained by the fact that the temperature conductivity of the heterogeneous system does not conform the additivity rule, i.e. the existence of foreign inclusions (in our case, fibrous FL) whose properties are radically different from the properties of the continuous medium (polymer binder) on the way of the heat flow leads to occurring of the additional heat energy source [12].

The coefficient of temperature conductivity *a* is a thermophysical property, which is a measure of the useful velocity of equating the temperature fields within the body [9], changes in an analogous way with the binder and OP based on it: it decreases with the temperature growth up to the area of the phase transition (Figure 1c), after which it starts increasing again. Increasing the content of Arimide-T fibre within the polyamide binder up to 15, 30 and 45 mass % provides for increasing the average temperature conductivity within the range of temperatures of 173-498 K by 11, 37 and 19 % correspondently. Increasing of *a* at introducing the fibrous FL can be explained by the increase of the bulk elasticity modulus and the growth of the useful length of the phonons' free path [12].

Thus, on the basis the held investigations, we can conclude that reinforcing PA-6 withArimide-Tpolyimide fibre allows to significantly improve thermophysical properties of OP and to expand the temperature interval of exploiting the goods from them.

REFERENCES

[1] Budnitsky G.A. Polymer fibres of the third generation: their development, properties, application // *Technical textile*. – №10. – 2004. – http://www.rustm.net.

[2] Limited liability company "Lirsot" // http://www.mytyshi.ru

[3] Burya A.I., Tkachenko E.V., Chigvintseva O.P. New organoplasticsbased on polyamides // *Composite materials*. – Volume 4. –№ 1. – 2010. – P. 89-96.

[4] Burya A.I., Tkachenko E.V. On the influence of arimide fibres on mechanic properties of aliphatic polyamide // Theses of reports of the III International Science and Technology Conference of students, post-graduates and young scientists *"Chemistry and modern technologies"*. – May 22-24, 2007. – Dniepropetrovsk. – P. 132.

[5] Askadskii A.A., Kondraschenko V.I. Computer material science of polymers. – Vol.1. – Atom-molecule level. M.: *Scientific world,* 1999. – 544 p.

[6] Lipatov Yu.S., Privalko V.P. Colorimetric investigation of filled linear polyuretanes // *High-molecular compounds*. – 1971. – A. – Vol. 13. – P.103-110.

[7] Lipatov Yu.S. Physical chemistry of filled polymers. M.: *Chemistry,* 1977.–P. 99.

[8] Solomko V.P., Pas'ko S.P. Investigation of the specific heat capacity of the filled polystyrene and polymethylmethacrylate // *High-molecular compounds*. – 1970. – Vol.(A)XII.– № 3. – P. 681-686.

[9] Thermophysical and rheological properties and the coefficients of friction of filled thermoplastics/ V.A. Paharenko, V.G. Zverlin, V.P. Privalko et al. *Reference book*. – Kiev: Naukova dumka, 1983. –279 p.

In: Chemical and Biochemical Kinetics: New Perspectives ISBN 978-1-61209-224-9
Editor: Gennady Zaikov

Chapter 22

IR-SPECTRA AND A CRYSTALLINE STRUCTURE OF PENTON AND THE CARBOPLASTIC BASED ON IT

A. S. Redchuk**[*]**, ***A. I. Burya and V.V. Golovyatinskaya***
Dneprepetrovsk State Agrarian University,
Dnepropetrovsk, Ukraine;

ABSTRACT

The features of the IR spectra for pressing Penton powder and carboplastic on its basis were studied. It is concluded that as the process of pressing, and the introduction of carbon fiber, change the structure of the polymer in the direction of increasing the content of amorphous phase. Shown that the pressed polymer mainly contains crystals of β-modification, whereas in the carboplastic present and β- and α-modification.

Keywords: carboplastic, Penton, IR-spectrum, the crystal structure

INTRODUCTION

Penton (poly-3, 3-bis (chloromethyl) oxacyclobutane) belongs to the polymers of a very high degree of crystallinity. The glassy state can be obtained only in case of the melt's cooling at low (about 0°C) temperatures, and at environ temperature it is transferred into the crystallineβ-modification [1]. β-form crystals are also obtained in case of the melts' crystallization at high (about 5000 atm) pressures. In case of the melts' crystallization at environ temperature α-modification crystals are formed.

Penton (poly-3, 3-bis (chloromethyl) oxacyclobutane) IR-spectrum is investigated rather completely [2-5].

The authors [2] investigated the melts and films of the polymer and stated that the absorption bands of 1027cm^{-1} and 897cm^{-1} belong to the amorphous and crystalline phases of

[*] E-mail: aredchuk@mail.ru.

polymers correspondingly. The band of $1027cm^{-1}$ shows quite a low intensity within the annealed films, while being intensive for melt and glass specimens, and there is no band of $897\ cm^{-1}$.

There is a thorough reference of absorption bands to the functional groups fluctuations by means of calculating the normal vibrations in [3]. The authors of this work refer absorption bands to three isomeric modifications of elementary units of polymers different by the location of chloromethyl groups.

In works [4, 5] we can find the investigation of the influence of polymorphic transformations (transfer of a less stable β-modifications at ordinary conditions into a more stable α-modification) on IR-spectrum of absorption. The authors have shown that such a transformation mainly influences the bands intensity within the area of $1300\text{-}1320cm^{-1}$.

The purpose of this study was a comparative investigation of the structure of Penton and the carboplastic based on it and Ural fiber by infrared spectroscopy.

METHODS OF INVESTIGATION

We obtained the spectra of initial Penton powder, pressed specimens and specimens of the composite containing 10 mass % of Ural fibre. The specimens of composites with higher fibre content could not be researched due to strong scattering of irradiation by the fibre. IR-spectra were registered at IKS-29 (LOMO, St-Petersburg) spectrophotometer within the range of $4000 - 400cm^{-1}$. The specimens were prepared as potassium bromide pellets, the pressed polymer and the composite were previously ground. The additions of KBr and the investigated materials made 600 and 30 mg correspondingly. The registered spectra were digitized, and the coefficient of transmission was translated into the absorbancy. Absorption bands were referred with the concern of data presented in [2-5]. Ural fibre taken in amounts introduced into the composite does not have any distinct absorption bands, which is why all the observed bands were referred to the polymer absorption.

INVESTIGATION RESULTS

The wave numbers of the absorption bands are cited in table 1. Within the area of hydrogen stretching modes in all the spectra we can observe a wide band with poorly seen peak at about $3400cm^{-1}$, which can be referred to the specimens' containing moisture. Besides, within the area below $3000cm^{-1}$ in all the spectra there is a group of intense mutually overlapping bands of symmetrical and antisymmetrical stretching modes of CH_2 groups of both methylene ones of the basic polymer chain and chloromethyl ones. The vibrations strongly interact and cannot be referred exactly.

Within the area of $2200 - 1500cm^{-1}$ there are some relatively weak bands which cannot be referred to Penton vibrations. As it is mentioned in [2] and as it is concluded from oxacyclobutan spectrum analysis [6], these bands should be referred to the remains of monomer in Penton.

The most structurally interesting Penton absorption bands are located in the area of $1500 - 400\ cm^{-1}$, and the corresponding spectra are presented at Figure 1.

Table 1. Wave numbers of absorption bands of IR-spectra of Penton and the carboplastic containing 10 % of Ural fibre

Wave numbers of the peaks			Intensity, shape	Reference
Penton powder	Pure pressed Penton	Carboplastic		
~ 3400	~ 3400	~ 3400	wide	moisture
2962	2977	2962	m- s	ν_s CH_2
2948	2965		w	ν CH_2
2899	2918	2901	s	ν_{as} CH_2
2883	2907	2888	s, overlap.	ν CH_2
2870	2900	2872		
	2891	2858		
2822	2835	2823	w, wide	?
2116	2122	2121	w, wide	Trace of monomer
1728	1727	1726	w	
1632	1629	1632	w	
1603	1594	1599	w, wide	?, possible moisture
1486	1482	1484	s	δ CH_2
1462	1460	1461	m	
1446	1440	1442	s	
	1429	1430	w	δ CH_2 with the share of ν CC
1426	1424	1425	s	
1384	1380	1383	s	
1323	1308	1313	s	
1270	1261	1268	s	δ CH_2
1105	1107	1102	vs, wide	
1071	1071	1066	s, wide	
	1052		w, shoulder	ν CO
1017	1019	1018	w, shoulder	ν CO
	916	913	w	δ CH_2
894	897	894	s	ν CC with the share of δ CH_2
862	863	860	s	
788	789	786	m	chain CC with the share of CCl
	754	751	m- s, overlap.	
743	745	743		
696	698	696	s	ν CCl
623	624	621	m	ν CCl with the share of ν CC
599	598	596	m	
523	523	520	m- s	
428	428	425	m	δ CCC

Notions: s – strong, m – medium, w – weak, v – very;ν - stretching modes, δ - bending modes.

The absence of significant absorption within the area of 1020cм$^{-1}$ and high intensity of 897 и 523cm^{-1} bands testifies to the fact that Penton is mainly crystalline in all the investigated specimens [2, 4]. The main differences between the spectra of the powder, pressed Penton and composite are the following.

AREA 1300 – 1500CM^{-1}

Besides the intensive absorption bands of crystalline phases, there are weak peaks in this area, which, according to [3, 4], can be considered as the vibrations of the polymer's amorphous phase. Thus, the band of 1460cm^{-1} within the powder spectrum is very weak, it significantly strengthens in the spectra of the pressed specimens, and gets rather intensive and definite within the composite spectrum. The band of 1430cm^{-1}, which is also referred to the amorphous phase, is not observed within the Penton powder spectrum and occurs in the spectra of the pressed Penton and the composite. The characteristics mentioned allow to suggest that the amorphous phase's content reaches the minimum value within the Penton powder, and it reaches its peak within the composite.

The authors [4] showed that during Penton's transition from β-modification into α-modification the absorption band of 1310cm^{-1} shifts to 1318cm^{-1}. Within the obtained powder spectrum we can observe a band of 1323cm^{-1}, while within the spectra of the pressed Penton and the composite this band is found at 1308 and 1313cm^{-1} correspondingly, and the latter shows a clearly widened band. Thus, we can presuppose that the pressed Penton shows the predominance of β- modification, while both shapes of crystals are represented within the composite.

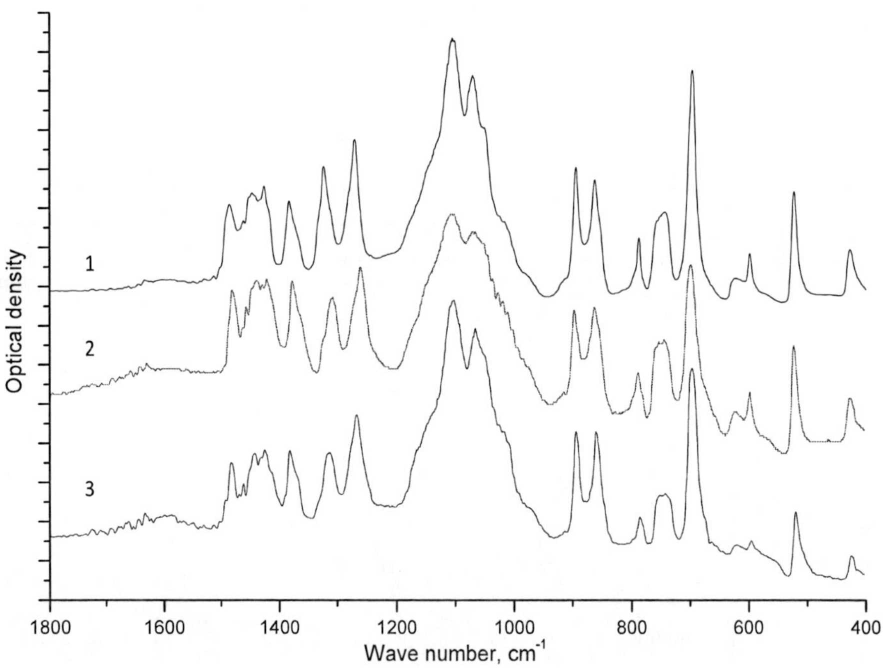

Figure 1. IR-spectra of Penton powder (1), pressed Penton (2) and carboplastic containing 10 % of Ural fiber (3).

AREA 1000 – 1100CM^{-1}

Within this area all the spectra show the set of some strongly overlapping intense absorption bands with the peaks at about 1105 and 1070cm^{-1}. At a short-wave wing of the last band we can notice a weak band of ~ 1020cm^{-1} which is referred to the amorphous phase absorption by the authors of all the previous investigations. Within the Penton powder spectrum this lever obviously shows a rather weak intensity, while within the pressed Penton spectrum it gets more distinct, and within the composite spectrum its intensity increases further, which confirms the previous conclusion on the increase of the amorphous phase's content in the row of "powder – pressed polymer – composite".

AREA 850 – 900CM^{-1}

Within this area one of the crystalline bands is located – 897cm^{-1}, the one which is absent within the spectra of melts and glass specimens and which possesses the high intensity within the Penton crystalline specimens' spectra. According to [4, 5], the intensity of the adjacent band (862cm^{-1}) is exposed to change to a much less extent in case of Penton films annealing. Within the powder spectra that we obtained, the intensity of the band of 894cm^{-1} is obviously higher than that of the band of 862cm^{-1}, while both bands possess almost equal intensity within the pressed Penton and composite spectra, which proves a lower degree of order within these specimens' structures.

CONCLUSIONS

The obtained data testify to the fact that the Penton powder pressing and especially introducing the filler into it results in the significant increase of the number of disordered areas within the polymer structure.

The investigation of the bands' shape and location within the area of 1300 – 1320cm^{-1} allows to suggest that the pressed Penton includes the polymer of the mainly less stable β-modification, while the crystalline areas of the carboplastic specimens include both polymer modifications.

REFERENCES

[1] Sandiford D.J.H. Polymers from oxacyclobutans.2. *J. Appl. Chem.* – v. 8. – 1958. – p. 188-196.

[2] Hatano M., Kambara S. Infrared spectra and crystallization of poly-3,3-bis(chloromethyl)oxacyclobutane. *J. Appl. Polymer Sci.* – v. VI. – N20. – 1962. – p. 232-239.

[3] Enomoto S., Opaskar C.G., Krimm S. Infrared spectra and structure of poly-3,3-bis(chloromethyl)oxacyclobutane. *J. Polymer Sci,* Part C. – N16. – 1967. – p. 2263-2276.

[4] Witenhafer D. E., Koenig J. L. Infrared studies of poly[3,3-bis(chloromethyl) oxacyclobutane]. I. Single crystals. *J. Macromol. Sci,* Part B. – v. 2. – N2. – 1968. – p. 235-246.

[5] Witenhafer D. E., Koenig J. L. Infrared studies of poly[3,3-bis(chloromethyl) oxacyclobutane]. II. Polymorphic transformation. *J. Macromol. Sci,* Part B. – v. 2. – N2. – 1968. – p. 247-260.

[6] *Spectral Database for Organic Compounds* (SDBSWeb): http://riodb01.ibase.aist.go.jp/sdbs/

In: Chemical and Biochemical Kinetics: New Perspectives ISBN 978-1-61209-224-9
Editor: Gennady Zaikov

Chapter 23

TEMPLATE SYNTHESIS OF BIOINORGANIC POLYMERS ON IONS OF TRANSITIVE METALS

Victor Kablov, Dmitry Kondrutsky* and Olga Zineeva
Volzhsky Polytechnic Institute (branch of) Volgograd State Technical University, Volzhsky, Russia

ABSTRACT

The researches in the field of bioinorganic polymers producing on the nickel, zinc and cobalt ions by template synthesis are carried out, and also selective sorption activity and termooxidizing destruction of these polymers are investigated. The synthesized template bioinorganic polymers are perspective for use as ecologically safe and economic sorption materials in processes of water purification, selective extraction of ions of nickel of technological solutions or sewage, in hydrometallurgy, and also as components fire-heat-shielding of materials.

Keywords: template synthesis, bioinorganic polymers, selective sorption materials, fire-heat-shielding of coverings

INTRODUCTION

All over the world there is a big interest to template synthesis as to a perspective method of reception new materials connections from more simple blocks more and more grows, reception which is complicated by traditional methods, and sometimes and at all it is impossible [1, 2].

In the present work results of research template synthesis of bioinorganic polymers on ions of transitive metals are presented. The chemical structure of bioinorganic polymers, sorption properties and thermal-oxidative degradation are investigated.

* E-mail: kondrutsky@gmail.com.

The urgency of work is commensurable with opportunities which are opened with use of the polymers received by template synthesis, as ecologically safe and economic sorption materials, and also as components fire-heat-shielding of coverings.

Experimental Part

We used as polymer food gelatin, as the supplier of an ion of metal – various salts ($NiCl_2 \cdot 6H_2O$, $ZnSO_2 \cdot 7H_2O$, $CoCl_2 \cdot 6H_2O$), as the sewing agent – formalin, and as forming porous agent – a hydrocarbonate of sodium.

The gelatin preliminary dissolved in water, mixed up with a solution of salt of nickel, zinc or cobalt. To this mix the solution of a hydrocarbonate of sodium was added. To the received complex of gelatin with salt and soda formalin was slowly flowed.

Synthesis was spent in soft conditions, in the water environment at temperature 30-40°C. Reaction proceeds with high speed, in a minute after mixture of all reagents the product is formed. The received polymer represents a macromolecular complex on ions of nickel (BIP-Ni), zinc (BIP-Zn) or cobalt (BIP-Co).

The bioinorganic matrix of comparison on gelatin BP-M Gelatin was received with use of the same reagents except for salt of metal.

For studying sorption properties of polymers was received a bioinorganic polymer-matrix (BIP-M) by extraction template ion of nickel by 2 % hydrochloric acid at intensive hashing in current of 20-30 minutes.

For studying properties of the received bioinorganic polymers was done thermal-oxidative degradation by a method of pyrolysis at presence of oxygen of air at high-temperature electric furnace SNOL.

Samples of bioinorganic polymers were preliminary weighed and were located in crucibles with known weight. Thermal-oxidative degradation was spent up to 500°C – the temperature value necessary for decomposition of an organic basis of polymers.

A SPECORD-M82 was used to directly characterize the polypeptides. IR spectrum was performed for the product in form of thin suspension in vaseline.

For researches of sorption propertiesa series of solutions of chloride of nickel with $C_н(Ni^{2+})$ = 0,05 mol/l, 0,01 mol/l and 0,005 mol/l and sulfate of copper with $C_н$ (Cu^{2+}) = 0,05 mol/l, 0,01 mol/l and 0,005 mol/l prepared. For sorption were used samples of water BIP-M and BP-M Gelatin. Samples of polymeric matrixes in weight 5 g were located in plastic glasses and were filled in with the prepared solutions of chloride of nickel and sulfate of copper. Later 5 day contents glasses was filtered and washed out by the distilled water. The received polymer was left on drying for definition sorption abilities BIP-M and BP-M Gelatin and calculation to the SEC.

Differential-thermal analysis (DTA) of product and gelatin was carried out from 25 to 600°C using a derivatograph "MOM", Hungary. The mass of samples was 120 mg. Rate of heating was 10^0C per minute. Sensitivity DTA was $1/5^0$C.

RESULTS AND DISCUSSION

IR Spectrometry of Products and Reagents

IR spectral research BIP-Ni, BIP-M, gelatin and BP-M gelatin has shown, that in spectra of substances there are no valent fluctuations of amin-groups to which wave numbers in the field of 3100 – 3400 sm^{-1} but which are present at a spectrum of gelatin.

Frequencies of absorption of IR in areas 1540 – 1650 sm^{-1}, 1360 – 1450sm^{-1}, 1075 – 1190 sm^{-1} corresponding carboxy group, are displaced in case of BIP-Ni and BIP-M in area with greater wave numbers that it is possible to explain their participation in formation of coordination communications on ions of nickel.

Strong strips of absorption in areas 1630 – 1680, 1550 sm^{-1} adequating to fluctuations aamide groups (Amide I and Amide II) also are displaced in case of BIP-Ni in more long-wave area that speaks processes intermolecular forming of complex.

In all substances the maintenance OH-groups, that shows in deformation fluctuations in the field of 850 – 960 sm^{-1}.

In BIP-Ni, BIP-M and BP-M Gelatin the maintenance – CH_2 – groups which source is formaldehyde increases, on what specifies increase in intensity of absorption of IR in the field of 1235 – 1255 sm^{-1}.

Comparison IR of spectra for polymers on ions Ni^{2+}, Zn^{2+} and Co^{2+} and on gelatin has shown matrixes of comparison, that each of three ions approximately to the same extent displaces waves carbox groups, amide groups in comparison with the polymer synthesized without participation oftemplate ion that show occuring processes of complex forming.

The Probable Mechanism of Reaction Template Synthesis

Studied reaction represents reception of a complex of gelatin and salts of transitive metal and the subsequent interaction of formaldehyde with this complex. Reaction goes to some stages.

At the first stage there is an interaction of gelatin with salt of transitive metal with formation of a complex of gelatin with an ion of metal:

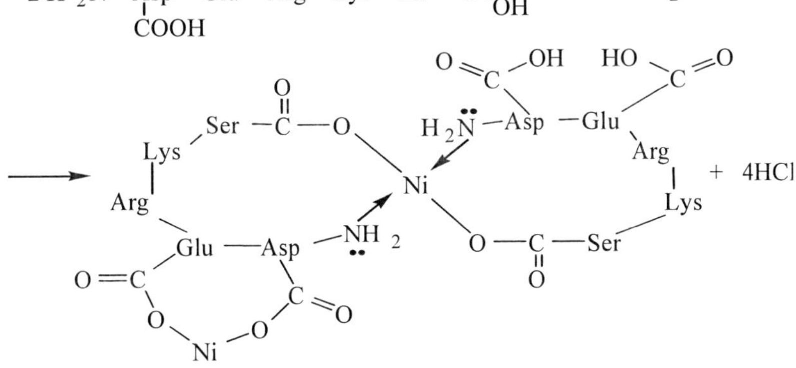

In the same time proceeds the reaction of forming porous with participation of a sodium hydrocarbonate.

At the second stage there isan attack of atom of carbon groups of formaldehyde in free electronic pair atom of nitrogen of a trailer amino group of gelatin:

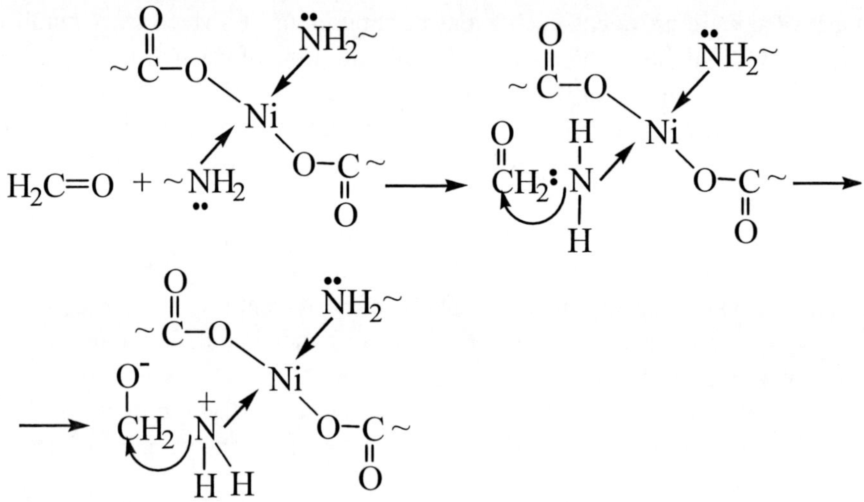

Formed intermediate complex attaches a free proton of water and turns in kation:

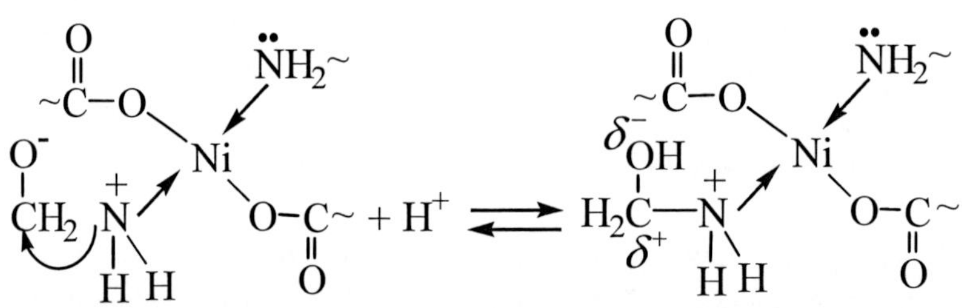

Molecules of polymer form the ordered structure around of an ion of metal which is sewed by formaldehyde bridges:

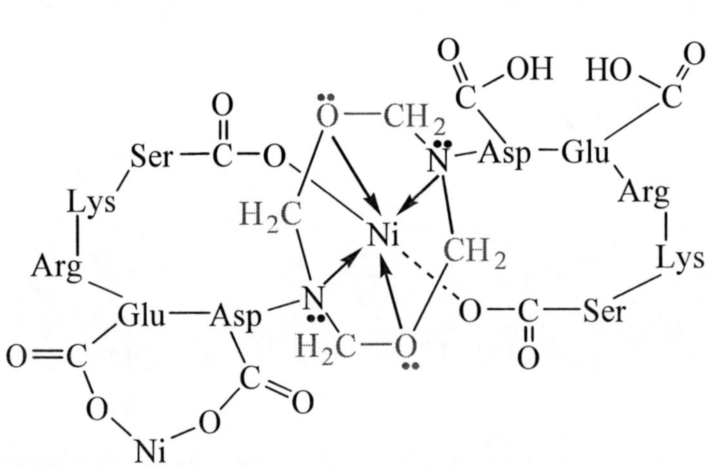

Research of Stability to Thermal-oxidative Degradation

A series of tests has allowed us to obtain data that has shown that introduction of an ion of metal as a matrix and structurization of polymer cross-section metilen bridges due to formaldehyde has led to increase in stability of polymer to thermal-oxidative degradation. Besides it has been proved, that ions of nickel possess greater structuring ability.

Research of Sorption Activity of Bioinorganic Polymers

Initial and final concentration of ions Ni^2 in investigated poly-measures pays off under the formula:

$$C(Ni) = \frac{C^{c.s.}(Ni^{2+}) \cdot V^{c.s.} \cdot V_{n2} \cdot V_{n1} \cdot V_{n0}}{V_{c1} \cdot \check{n}_{samp}}, mol/l \tag{1}$$

where $\acute{N}^{\acute{n}.s.}(Ni^{2+})$ – the concentration of ions of nickel from calibre schedule, mg/l; $V^{c.s.}$ – the volume of test of a solution for photocolorimeter analysis, ml; V_{n0} – the volume of a flask for preparation of a solution on sorption (500 ml); V_{n1} – the volume of the first test of cultivation, ml; V_{n2} – the volume of the second test of cultivation, ml; V_{c1} – volume of a flask with an analyzed solution on FEC (50 ml).

The static exchange capacity pays off under the formula [3]:

$$SEC = C_F(Ni) - C_0(Ni), \; mg\text{-}eq/g \tag{2}$$

where $C_F(Ni)$ – concentration of ions of nickel in a matrix after sorption; $C_0(Ni)$ – concentration of ions of nickel in a matrix before sorption.

The factor of distribution:

$$K_d = \frac{SEC}{C_s}, \tag{3}$$

where $C_s(Ni^{2+})$ – the concentration of sorption solution.

Results of experiments and calculations are presented in Tables 1 – 4.

The static exchange capacity in case of sorption BP-M Gelatin in a solution of sulfate of copper pays off under the formula [4]:

$$SEC_{BIP-M}(Cu) = \frac{C_{н}^{0}(Cu) - C_{н}^{F}(Cu)}{m_{sorp.dry}} \quad (4)$$

The generalized data on sorption investigated polymers in solutions of "related" and "another's" ions are presented in picture 1.

Table 1. Static exchange capacity and factor of distribution for BIP-M in a solution of chloride of nickel

№	$C_0(Ni^{2+})$, mg-eq/g	$C_s(Ni^{2+})$, mol/l	$C_F(Ni^{2+})$, mg-eq/g	SEC (Ni^{2+}), mg-eq/g	% absorption (Ni^{2+})	$K_d(Ni^{2+})$
1	0,07	0,0518	0,54	0,47	98	9,1
2		0,0097	0,22	0,15	40	15,5
3		0,0053	0,15	0,08	27	15,1

Table 2. Static exchange capacity and factor of distribution for BP-M Gelatin in a solution of chloride of nickel

№	$C_s(Ni^{2+})$, mol/l	SEC (Ni^{2+}), mg-eq/g	$K_d(Ni^{2+})$
1	0,0801	0,36	4,5
2	0,0102	0,14	13,7
3	0,0046	0,07	15,2

Table 3. Static exchange capacity and factor of distribution for BIP-M in a solution of sulfate of copper

№	$C^0_s(Cu^{2+})$, mol/l	$C^f_s(Cu^{2+})$, mol/l	SEC (Cu^{2+}), mg-eq/g	$K_d(Cu^{2+})$
1	0,0510	0,044	0,40	7,8
2	0,0255	0,0209	0,12	12,6
3	0,0116	0,0093	0,07	13,9

Table 4. Static exchange capacity and factor of distribution for BP-M Gelatin in a solution of sulfate of copper

№	$C^0_s(Cu^{2+})$, mol/l	$C^f_s(Cu^{2+})$, mol/l	SEC (Cu^{2+}), mg-eq/g	$K_d(Cu^{2+})$
1	0,0510	0,044	0,787	15,4
2	0,0255	0,0209	0,510	20,0
3	0,0116	0,0093	0,254	21,9

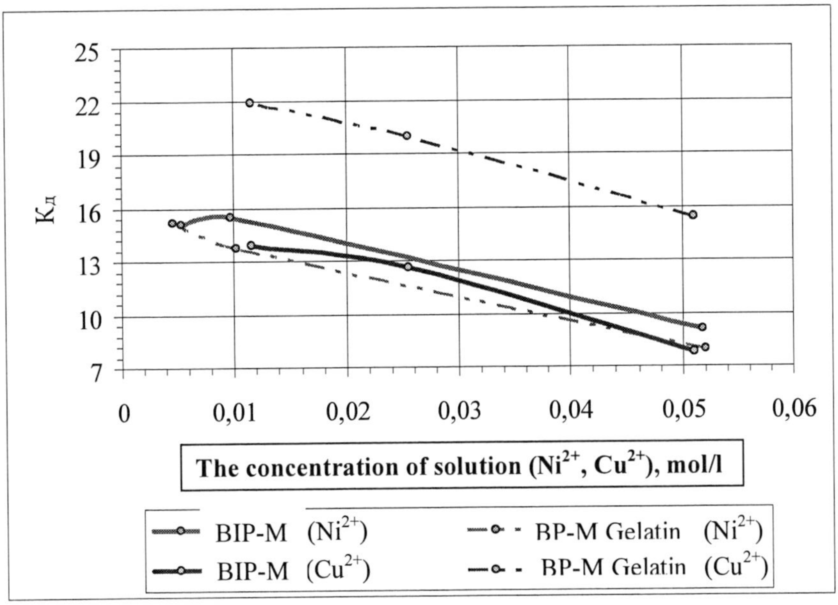

Picture 1. Dependence of factor of distribution on concentration of solutions Ni^{2+} и Cu^{2+} for BIP-M and BP-Gelatin.

Thus, according to the purposes it is proved, that template synthesis leads to reception of the materials possessing selective properties in relation to template ion. The bioinorganic matrix of polymer (BIP-M) synthesized on ionsof nickel in a series of tests proves as template sorbent possessing high absorbing selectivity to related ions and low to another's. So, factors of distribution for BIP-M in solutions with concentration 0,01 mol/l are K_d (Ni^{2+}) =15,5 and K_d (Cu^{2+}) =12,6.

It is proved, that the bioinorganic matrix of comparison on gelatin (BP-M Gelatin) possesses opposite selective properties. This matrix has been synthesized without template ions of metal, and it does not possess properties of template memory.

Differential Thermal Analysis and Thermo Gravimetric of Polymers and Gelatin

Results of research thermal degradation ofBIP-Ni, BIP-Zn and BIP-Co and initial gelatin by method of differential thermal analysis have shown that introduction of template ion raises stability of polymer to thermal-oxidative degradation.

During heating materials up to 100°C begins allocation in a gas phase adsorbe waters with expenses energy.

Up to 300°C processes of structurization of materials proceed. There is an allocation flying vapour waters, low-molecular flying substances due to formation of amide. This process is displayed in the form of peaks on curves DTA for bioinorganic polymers. Gelatin is exposed significant destruction with allocation of flying substances that is displayed in the form of peak on curve DTA.

After 250-300°C on curve sharp loss of weight by samples is clearly visible.

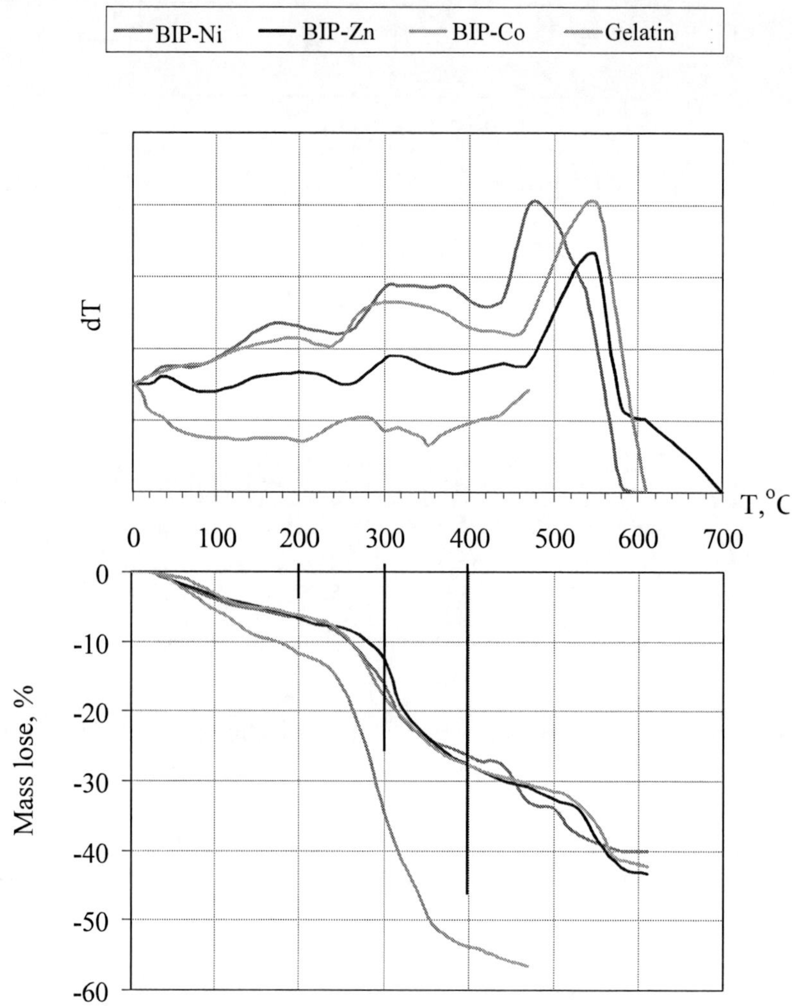

Picture 2. DTA and TG curves for BIP-Ni, BIP-Zn and BIP-Co and gelatin.

Thus, the received polymers have limits of thermostability 250-300°C. Thermostability of similar materials is caused by structuring influence of ions of metals which interfere with allocation of flying substances.

CONCLUSIONS

On the basis of the lead experiments and calculations is perspective using of the synthesized template bioinorganic polymers as ecologically safe and economic sorption materials in processes of water purification, selective extraction of ions of nickel of technological solutions or sewage, in hydrometallurgy, and also as components fire-heat-shielding of materials.

REFERENCES

[1] Pat.: CN101332999 (A) C01G3/02 / Method for preparing Cu_2O or CuO hollow submicrospheres with particle diameter controllable by water phase soft template method/ YUN FANG [CN]; JIN HU [CN]; YUEPING REN [CN]; YONGMEI XIA [CN].

[2] Ion exchange activity of the crosslink calixarene polymer produced by template synthesison base ofNa^+, K^+ и Ba^{2+} matrixes/ G. N. Altshuler, L.P. Abramova, N.V. Malyshenko, G.U. Shkurenko, E.V. Ostapova // Proceedings of Russian Academy of Science. Chemical Series. 2005 – №8. – p. 1919.

[3] Synthesis and properties of phosphorus-containing polypeptides for effective ion-exchange materials of improved capacity / Victor Kablov, Dmitry Kondrutsky, Maria Sudnitsina// Progress in Monomers, Oligomers, Polymers, Composites and Nanocomposites / Nova Science Publishers. – New York. – 2009. – P. 113-123.

[4] The characteristics of phosphoprous-containing polypeptides synthesis and their properties / Victor F. Kablov, Dmitry A. Kondrutsky, Maria V. Sudnitsina // Proceedings of Volgograd State Technical University. –Volgograd. 2008. – (Ser. Chemistry and technology hetero-organic monomers and polymer materials. Iss. 5)P. 121-125.

INDEX

A

B

C

D

E

F

G

H

Q

R

S

T

U

V

W

X

Y

Z